Futuristic Manufacturing

Increased industrial capacity, manufacturing output, and manufacturing technology all contribute significantly to a country's GDP. Manufacturing is the foundation of industrial production, so improving its methods and infrastructure is crucial for progress. Recent years have seen the introduction of a wide range of energy- and resource-efficient, environment-friendly, and occupationally safe manufacturing techniques, and this book focuses on these latest techniques, as well as continuous advancement, in order to meet current challenges. The book is divided into three sections: (1) subtractive manufacturing, (2) additive manufacturing, and (3) the use of artificial intelligence in manufacturing. It discusses micromachining, metal-based additive manufacturing, polymer-based additive manufacturing, hybrid additive manufacturing, and finally artificial intelligence in manufacturing.

Futuristic Manufacturing: Perpetual Advancement and Research Challenges connects modern manufacturing methods and emerging trends in the industry. It adds a thorough examination of modern manufacturing techniques and modifications that may be implemented in the future, and is an excellent resource of information for undergraduate and graduate students in manufacturing.

Science, Technology, and Management Series

Series Editor: J. Paulo Davim, *Professor, Department of Mechanical Engineering, University of Aveiro, Portugal*

This book series focuses on special volumes from conferences, workshops, and symposiums, as well as volumes on topics of current interested in all aspects of science, technology, and management. The series will discuss topics such as, mathematics, chemistry, physics, materials science, nanosciences, sustainability science, computational sciences, mechanical engineering, industrial engineering, manufacturing engineering, mechatronics engineering, electrical engineering, systems engineering, biomedical engineering, management sciences, economical science, human resource management, social sciences, engineering education, etc. The books will present principles, models techniques, methodologies, and applications of science, technology and management.

Multi-Criteria Decision Modelling
Applicational Techniques and Case Studies
Edited by Rahul Sindhwani, Punj Lata Singh, Bhawna Kumar, Varinder Kumar Mittal, and J. Paulo Davim

High-k Materials in Multi-Gate FET Devices
Edited by Shubham Tayal, Parveen Singla, and J. Paulo Davim

Advanced Materials and Manufacturing Processes
Edited by Amar Patnaik, Malay Kumar, Ernst Kozeschnik, Albano Cavaleiro, J. Paulo Davim, and Vikas Kukshal

Computational Technologies in Materials Science
Edited by Shubham Tayal, Parveen Singla, Ashutosh Nandi, and J. Paulo Davim

Industry 4.0 and Climate Change
Edited by Rajeev Agrawal, J. Paulo Davim, Maria L.R. Varela and Monica Sharma

Sustainable Materials and Manufacturing Technologies
Edited by Navneet Khanna, Kishor Kumar Gajrani, Khaled Giasin and J. Paulo Davim

Futuristic Manufacturing
Perpetual Advancement and Research Challenges
Edited by Mithilesh K. Dikshit, Vimal Kumar Pathak, Asit Baran Puri and J. Paulo Davim

For more information about this series, please visit: www.routledge.com/Science-Technology-and-Management/book-series/CRCSCITECMAN

Futuristic Manufacturing

Perpetual Advancement and Research Challenges

Edited by Mithilesh K. Dikshit, Vimal Kumar
Pathak, Asit Baran Puri and J. Paulo Davim

CRC Press
Taylor & Francis Group
Boca Raton London New York Leiden

CRC Press is an imprint of the
Taylor & Francis Group, an **informa** business

A BALKEMA BOOK

Designed cover image: Mithilesh K. Dikshit

First published 2023
by CRC Press/Balkema
4 Park Square, Milton Park, Abingdon, Oxon, OX14 4RN
e-mail: enquiries@taylorandfrancis.com
www.routledge.com—www.taylorandfrancis.com

CRC Press/Balkema is an imprint of the Taylor & Francis Group, an informa business

Library of Congress Cataloging-in-Publication Data
A catalog record for this book has been requested

ISBN: 978-1-032-21779-6 (hbk)
ISBN: 978-1-032-21790-1 (pbk)
ISBN: 978-1-003-27002-7 (ebk)

DOI: 10.1201/9781003270027

Typeset in Times New Roman
by Apex CoVantage, LLC

Contents

About the Editors

Mithilesh K. Dikshit is Assistant Professor at Institute of Infrastructure Technology Research and Management (IITRAM), India. He has a Ph.D. in mechanical engineering and a master's degree in production engineering. He has more than 11 years of teaching and research experience. High-speed machining, metal cutting mechanics, composite materials, molecular dynamics, vibration and control, and optimization are among his major research interests. He has organized a number of conferences and symposiums. He has over 35 research papers and journal articles to his credit.

Vimal Kumar Pathak earned his Ph.D. in mechanical engineering (design) from MNIT Jaipur in Rajasthan, India. He earned his M.Tech. from ISM Dhanbad in India. He is currently employed as Assistant Professor in the Department of Mechanical Engineering at Manipal University Jaipur in Jaipur. He is currently working on the development of novel hybrid optimization algorithms and their applications. His research interests include optimization, soft computing, composites, metrology, and additive manufacturing. He has over 50 journal and conference publications to his name.

Asit Baran Puri is Professor at the National Institute of Technology Durgapur in the Department of Mechanical Engineering. He received a first-class degree in mechanical engineering from the National Institute of Technology Durgapur (formerly R.E. College Durgapur) in West Bengal, India, in 1990. He earned his master's degree in 1992 and his Ph.D. in 2003 from Jadavpur University in Kolkata, India. He has over 70 research papers published in prestigious international journals and conference proceedings. Non-conventional machining, metal cutting theory, machine tool engineering, micromachining, and process optimization are among his research interests.

J. Paulo Davim is a mechanical engineering Professor at the University of Aveiro in Portugal. He holds a Doctor of Science degree from London Metropolitan University as well as a Ph.D. in mechanical engineering from FEUP University of Porto. He has more than 30 years of teaching and research experience in manufacturing, materials, mechanical, and industrial engineering, with a focus on machining and tribology. He has served as the chief editor of several prestigious international journals.

Contributors

Acherjee Bappa, Department of Production and Industrial Engineering, Birla Institute of Technology Mesra, Ranchi, India

Alam Md. Quamar, Department of Mechanical Engineering, Indian Institute of Technology Patna, India

Chakrabarti Debalay, IIT Kharagpur, India

Chattopadhyaya Somnath, Department of Mechanical Engineering., Indian Institute of Technology (ISM), Dhanbad, India

Das Alok Kumar, Department of Mechanical Engineering, Indian Institute of Technology (ISM), Dhanbad, India

Das Apurba, Department of Aerospace and Applied Mechanics, IIEST, Howrah, India

Duggirala Aparna, School of Laser Science and Engineering, Jadavpur University, Kolkata, India

Elangovan S., PSG College of Technology, Coimbatore, India

Khan Dilshad Ahmad, Department of Mechanical Engineering, National Institute of Technology Hamirpur, India

Kumar S. Pratheesh, PSG College of Technology, Coimbatore, India

Kumar V., Indian Institute of Technology (ISM) Dhanbad, India

Kumar Vidyapati, IIT Kharagpur, India

Kundu Anupam, IIT Kharagpur, India

Mandal A., IIT Kharagpur, India

Mitra Souren, Department of Production Engineering, Jadavpur University, Kolkata, India

Mohanraj R., PSG College of Technology, Coimbatore, India

Muralidhar Manapuram, NERIST, Itanagar, India

Ozah Rupshree, NERIST, Itanagar, India

Pratihar D.K., IIT Kharagpur, India

Puri Asit Baran, National Institute of Technology Durgapur, India

Rahman Md. Zishanur, Department of Mechanical Engineering, Nalanda College of Engineering, Nalanda, India

Rajamani R., PSG College of Technology, Coimbatore, India

Ramshankar C.S., Maxbyte Technologies Private Limited, Coimbatore, India

Rana Aarti, Department of Mechanical Engineering, National Institute of Technology Hamirpur, India

Ray Debajyoti, Sanaka Educational Trust's Group of Institutions, Durgapur, India

Roy B.K., Indian Institute of Technology (ISM) Dhanbad, India

Sharma Arun, Department of Mechanical Engineering, National Institute of Technology Hamirpur, India

Sreekanth T.G., PSG College of Technology, Coimbatore, India

Sugumar R., Maxbyte Technologies Private Limited, Coimbatore, India

Vidya Shrikant, School of Mechanical Engineering, Galgotias University, Greater Noida, India

Wazeer Adil, School of Laser Science and Engineering, Jadavpur University, Kolkata, India

Preface

Manufacturing processes are the most important practices in industries to transform raw materials into useful products for the service of society and mankind. With the rapid development in materials engineering, several high-end materials with superior mechanical and physical properties are being developed for various applications. Processing these materials to get a useful product is a challenging task. Moreover, conventional manufacturing processes are not adequate to meet future challenges. Thus, there is a need to upgrade conventional manufacturing systems to meet future challenges with lower manufacturing costs. The objective of the book is to present the technological advancements in the manufacturing processes and research challenges. This book focuses on machining processes such as micromachining, sustainable manufacturing, hybrid machining processes, and additive manufacturing processes for metal, polymer, and hybrid additive manufacturing, and the application of artificial intelligence in manufacturing.

Each chapter was subjected to an evaluation by a panel of anonymous specialists in the relevant study field. The chapters were chosen based on their overall quality as well as their significance to the goals and objectives of the book. This book will function as a link between the manufacturing techniques that are used today and the manufacturing methods that will be used in the future. The reader will walk away with a comprehensive knowledge of the processes involved in advanced manufacturing, as well as a grasp of the possibility for modifications in subsequent implementations.

Mithilesh K. Dikshit
Vimal Kumar Pathak
Asit Baran Puri
J. Paulo Davim

Preface

Acknowledgments

This book is the outcome of the efforts of many people. It includes all the authors, reviewers, and the concerned institutes of editors. First, the editors would like to acknowledge all whose work, research, and support have helped and contributed to this book. The accomplishment of this book represents the collaborative efforts of a large number of people. It is vital that their role in sculpting the overall structure of the work be recognized and acknowledged. The editors would like to express their gratitude to each and every reviewer for the significant contributions they made in elevating the overall quality of the chapters. In conclusion, we would like to express our gratitude to everyone who has contributed to the chapters, both for their hard work and their perseverance.

It is not an easy task to find the right words to express appreciation to each and every person who made a contribution to this book. Due to this fact, we would like to take this opportunity to extend our gratitude to each and every person who contributed in any way, shape, or form toward the successful completion of the book. In addition, we would like to take this opportunity to express our gratitude to our friends and family members for all of the assistance and support they provided to us while we were working on this book.

Chapter 1

Mechanical Micro-milling

A Broader Perspective of the Process

Debajyoti Ray and Asit Baran Puri

1.1 INTRODUCTION

Miniaturization of the products and their manufacturing processes are considered to be key trends in the development of technology. At present, huge demand has emerged for the miniature components with complex micro-features. Micro-components have widespread applications in the areas ranging from medical science and implant technology, biotechnology, automotive and transport, aerospace to consumer products, electronics, etc. Micromachining processes contribute a larger proportion in the entire range of the micro-manufacturing activities. Silicon-based micro-components related to microelectromechanical systems (MEMS) and microelectronics have been fabricated by processes commonly known as lithography-based micro-manufacturing processes. The common methods include photo lithography, chemical etching, plasma etching, and LIGA. The micro-components made through these processes are essentially planar geometries (Chae et al., 2006). However, in the past 25 years or so, there has been an emerging requirement of the micro-components in commercial, healthcare, and defense sectors where the components were to be made from high strength materials having precise and complex features. To cater the increasing requirement of these free-form and high-accuracy micro-components needed in different application areas, there has been a growing need for reliable, flexible, and cost-effective methods to fabricate these components. Micro-manufacturing techniques such as laser beam machining, ultrasonic machining, ion beam machining, micro-EDM, and micro-ECM are slow processes and are limited in their applicability, as these processes can be carried out only on a few types of materials (Masuzawa and Tonshoff, 1997; Masuzawa, 2000). Mechanical micromachining or micro-cutting represents fast, direct, reliable, flexible, and economical micromachining process capable of fabricating complex-shaped 3D micro-products made of various types of materials such as metals, composites, polymers, and ceramics. Mechanical micromachining or micro-cutting is a cost-effective technique for mass production as well as small and medium lot size production of micro-parts. Micro-cutting refers to the material removal and creating machined features in microscales with no constraint on the size of the component being machined. Micro-cutting includes micro-turning, micro-milling, micro-drilling, and micro-grinding. Among all these micro-cutting processes, micro-milling process has the potential to create intricate micro-features and achieve form accuracy in shaping 3D micro-parts. Mechanical micro-end milling is a powerful and versatile process capable of creating micro-features in a wide variety of materials (Chae et al., 2006). Figure 1.1 schematically illustrates some micro-features created through mechanical micromachining processes. The features shown are micro-column and micro-channels fabricated by micro-milling. The

DOI: 10.1201/9781003270027-1

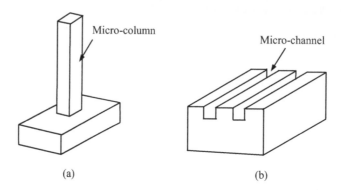

Figure 1.1 Schematic of micro-features: (a) micro-column and (b) micro-channels fabricated by micro-milling process.

dimensions of the cross section of these micro-features are in micrometer scale. The market need for micro-products with smaller features and tight tolerances created the scope for the development of micro-cutting technology. The researchers have gained opportunity to explore exciting challenges in fabrication of microscale devices by mechanical micro-cutting (Dornfield et al., 2006; Alting et al., 2003; Kussul et al., 1996).

Mechanical micro-milling may be described as cutting in micro-scale having similar kinematics to the conventional or macro-milling process. Both macro- and micro-mechanical cutting are characterized by mechanical interaction of the cutting edge of the cutting tool with the work material causing plastic deformation inside the material and removal of the material in the form of chips to produce the desired shape. However, there are a number of issues related to the chip removal mechanism and cutting process mechanics that arise in the micro-domain and influence the cutting process. The important aspects of cutting process mechanics are related with the material removal and the minimum chip thickness effect associated with the cutting-edge radius, low cut size, reduced workpiece imperfections, and increased non-homogeneity of the work material, highly negative effective rake angle encountered by the material layer to be cut, ploughing, "size effect" at reduced material removal, etc. All these factors influence the cutting process performances.

1.2   MICRO-MILLING PROCESS MECHANICS

Mechanical micromachining exhibits different characteristics compared to conventional macro-machining due to reduction of all the relevant dimensions in the machined features, cutting tool, and the cutting parameters. The size and geometry of the micro-milling tool determines the size and shape of the micro-features. Encompassing a wide range of views by various researchers, micro-milling may be defined as the micromachining process where the uncut chip thickness is comparable to the cutter-edge radius in size and uses small-diameter end mills (diameter less than 1,000 µm) creating feature sizes ranging from few micrometers to a few millimeters (Aramcharoen et al., 2008). Commonly, in micro-features, the size is below 1 mm (generally in the range of 1–100 µm) in at least two dimensions (Huo and Cheng, 2013). The diameter of the customized or the commercially available micro-end milling tools usually ranges from 50 μm to 1 mm having one or more cutting edges with or without helix angle. Low stiffness and high fragility of the micro-milling tool impose limitations on the operating values of the cutting variables such as feed rate and depth of cut. To avoid excessive chip load on the cutting edges and breakage of the micro-tools, the operating values of the feed rate and depth of cut are often kept at low values. Uncut chip thickness less than tens of micrometers are often used in the micro-milling process. Figure 1.2 depicts the comparison of uncut chip thickness with respect to cutting-edge radius in macro- and micro-cutting. In micro-cutting, downsized tool–work interaction due to low feed rate and uncut chip thickness significantly influences the micro-cutting mechanics and the process responses.

In micro-cutting, the uncut material thickness is also often comparable to the grain size or short-range atomic orders in the work material. Material removal largely takes place in front of the rounded cutting edge and the work material flow around the cutting edge needs to be considered to analyze the cutting mechanics. Therefore, it is important to analyze the different process characteristics in micro-cutting using a suitable micro-cutting model. In micro-cutting, at low uncut chip thickness or low depth of cut (orthogonal planning process), the cutting-edge radius influences the cutting process as the effective rake angle is highly negative. The work material below a certain level of uncut chip thickness (or depth of cut) is significantly ploughed at the rounded cutting edge. In micro-cutting, the ploughing action is dominant in the overall cutting action compared to macroscale cutting. Several studies on the behavior of the material flow around the cutting-tool edge that considered ploughing during cutting indicated that the approaching work material toward the rounded cutting edge bifurcates at some separation point near the cutting-tool edge. The inflowing material at a depth above the level of this point joins to the flowing out chip, whereas the lower layer below the separation point is pushed beneath the cutting edge to become part of the machined work material (Albrecht, 1960; Connolly and Rubenstein, 1968). This separation point may be described as a stagnation point having a location angle α_s, as shown in Figure 1.3. However, a slightly separate approach on the mechanism of ploughing process proposed by some researchers (Palmer and Yeo, 1963) indicated that the lower layer of the

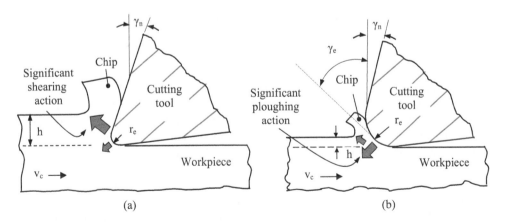

Figure 1.2 (a) Macroscale cutting, uncut chip thickness (*h*) much larger than cutting-edge radius (r_e).
(b) Microscale cutting, uncut chip thickness (*h*) much comparable to cutting-edge radius (r_e).

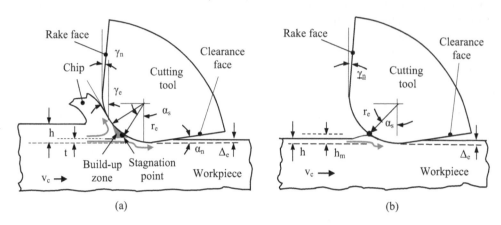

Figure 1.3 Schematic of the work material flow around the rounded cutting edge in micro-cutting when (a) $h > h_m$ and (b) $h < h_m$.

uncut material flow confronts the rounded cutting edge at a point where it gets locked up in front of the cutting edge due to the effective rake angle that is highly negative. As a result of this, the locked-up work material hardly finds way to pass away with the chip and forms a very small "buildup" adhering to the front of the cutting edge as a stable dead metal zone. Inflowing material toward the rounded cutting edge gets bifurcated at the tip of this small buildup zone where the upper layer diverges toward the chip and the lower layer of thickness t (Figure 1.3a) is ploughed and pushed below the cutting edge to move beneath the clearance face of the cutting tool. The depth of the material layer that is subjected to ploughing may be referred as the ploughed thickness. Though different physical scenarios may be thought of related to the material diversion at the cutting-tool edge, the tip of the buildup zone and the stagnation point may be considered synonymous. The tiny stable buildup metal zone and the corresponding stagnation point in cutting are shown in Figure 1.3.

1.2.1   Cutting-Tool Edge Radius Effect in Micro-cutting

In macroscale cutting, the uncut chip thickness is usually much greater than the cutting-tool edge radius. For example, in macro-milling with a 5-mm diameter end mill, the uncut chip thickness can be 0.3 mm, whereas the cutter-edge radius can be as small as 0.003 mm. Therefore, in macro-cutting, for a cutting tool with a positive nominal rake angle, the effective rake angle at the average uncut layer thickness is almost equal to the nominal rake angle. In this case, the rounded arc segment of cutting edge can be neglected and the straight rake face confronts the majority of the uncut layer. Larger proportion of the forces involved in macro-cutting may be attributed to the shearing of the work material forming the chip. Contrary to macroscale cutting, a different scenario exists in micro-cutting. In a typical micro-milling process with a 0.3-mm diameter micro-end mill, the maximum uncut chip thickness can be as small as 0.001 mm where the cutter-edge radius can be 0.003 mm. The cutting tool in micro-cutting process may have a positive nominal rake angle, whereas the effective rake angle at the average uncut layer thickness is largely negative. In microscale cutting, when the uncut chip thickness h exceeds h^*, which is equal to $r_e(1+\sin\gamma_n)$ as shown in Figure

1.4, the cutting scenario is shear dominant and the approaching uncut material confronts the cutting tool with edge radius r_e and nominal rake angle γ_n. When h is less than h^*, the incoming uncut material confronts the rounded edge of the cutting tool at effective rake angle (γ_e), which is negative corresponding to the cutting depth h, and is equal to $-\sin^{-1}\left(1-\dfrac{h}{r_e}\right)$.

h^* is the maximum uncut layer confronted by the rounded cutting edge in Figure 1.4. The incoming work material flow in ploughing dominant cutting scenario is restricted due to the barrier of the highly negative rake angle effective at the cutting point. When the uncut chip thickness (h) with respect to the cutting-edge radius (r_e) is very low, chip formation hardly takes place and the material is ploughed and pushed beneath the cutting edge. The ploughed material undergoes elastic–plastic deformation and flows beneath the flank face of the cutting tool with possible elastic recovery. Thus, the cutting-edge radius plays a significant role in the material removal mechanism. As mentioned earlier, in micro-cutting, the upper layer (of the total uncut layer) of the work material above a critical thickness rises up the rounded edge to form the chip, whereas the bottom layer of the incoming material gets ploughed by the rounded cutting edge. The proportion of ploughing and rubbing forces in the overall cutting action of the micro-cutting process is not insignificant. The involvement of the flat rake face gets reduced in micro-cutting and the interaction of the work material with the cutting-edge and the flank surface becomes significant. A larger proportion of the cutting energy is utilized in ploughing without contributing in material removal, giving rise to the size effect. Several studies in precision machining and micromachining have been done for understanding the influence of the cutting-edge radius in machining.

The effect of ploughing in micro-cutting is due to the pressure applied on the work material by the rounded cutting edge. Understanding of the micro-cutting process stemmed from the analysis of the ultraprecision cutting where the cutting depths of the material layer were also very small (i.e., in the range of 0.01–10 µm) and round-edge cutting models were adopted for the analysis of the cutting mechanics. In an analysis of the cutting forces that considered segregation of the ploughing force from the tool rake face force in precision cutting, it was shown that ploughing was associated with the movement of the uncut material ahead of the rounded edge mostly compressed into the machined surface. The amount of material sheared

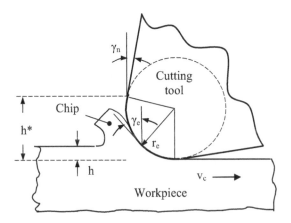

Figure 1.4 Schematic of the effective rake angle (γ_e) in microscale cutting.

was found to depend essentially on the uncut chip thickness and the amount of the ploughed material was found to depend on the magnitude of the sharpness radius of the cutting tool (Kim and Kim, 1995). Analysis of the precision cutting processes considered ploughing of the work material flowing below the incipient small buildup assumed to exist ahead of the cutting edge and the material eventually moving through the plastic region that existed beneath the lower curved surface of the rounded edge. These analyses revealed that cutting forces were influenced by the cutting-edge radius. The specific cutting energy and the specific thrust resistance was found to depend on the ratio of uncut thickness of cut to the cutting-edge radius for low uncut material layer comparable to the edge radius of the apparently sharp precision cutting tool (Abdelmoneim, 1980). Experimental analysis showed that in cutting with a large-edge radius cutting tool having feed rate below the cutting-edge radius, a significant amount of side burrs were formed and large ploughing forces compared to shearing forces were developed (Waldorf et al., 1999). Micro-cutting force models take into account the significant presence of the ploughing force component due to finite edge radius of the cutting tool comparable to the uncut chip thickness. Finite element analysis was used in micro-cutting models to estimate the stress distribution at the cutting edges and analyze the influence of the cutting-edge radius on the cutting forces (Wu et al., 2016). Nanometric cutting of brittle materials was studied through molecular dynamics simulation using cutting tools having different edge radii and varying depths of cut. A study using indentation sliding model revealed significant influence of the cutting-tool geometry and depth of cut on the variation of the cutting and thrust forces, specific cutting energy, and subsurface deformation (Komanduri et al., 1998). Studies on ultra-precision machining using single-crystal diamond showed that cutting-tool edge geometry has significant influence on the cutting forces and specific energy. These studies showed dominance of ploughing at low uncut thicknesses and increased cutting forces for increased negative rake angle. For uncut chip thickness larger than the cutting-edge radius, the resulting cutting forces were governed by the nominal rake angle of the cutting tool. However, as the uncut chip thickness approached the edge radius of the cutting tool, the effective rake angle significantly influenced the cutting forces. A study on cutting and thrust forces and specific cutting energy using new and worn-out tools revealed that the worn-out tools resulted in increased cutting and thrust forces than the new tools. Though both type of tools revealed size effect at low uncut chip thickness, the rate of increase in specific cutting energy with the decrease of uncut chip thickness is more in case of worn tool than the sharp edged new tool (Lucca and Seo, 1993). The energy dissipation and the stress–strain distribution in nanoscale cutting can be described using an atomic model based on the simulation studies. A study on nanoscale cutting revealed the existence of an area of high shear strain and high stress around the tool tip and along the rake face. The results obtained in the study showed that the energy dissipated in the surface generation remained small. However, the rate of energy dissipation in plastic deformation under the cutting edge was large in nanoscale cutting compared to that observed in macroscale cutting (Inamura et al., 1993).

1.2.2 Minimum Chip Thickness Effect in Micro-cutting

The uncut layer thickness in micro-cutting at very low feed rates can be so low that the inflowing work material hardly finds a way to rise over the rake face to be removed as chip. There exists a critical uncut thickness or removable material layer referred as minimum uncut chip thickness (MUCT) that determines whether a chip will form or not. This limiting uncut chip thickness or the removable material layer is referred as the minimum uncut chip thickness (h_m). For a certain low value of uncut chip thickness (h) with respect to edge radius of tool (r_e), material is not removed in the form of chip as illustrated in Figure 1.5. Difficulty arises in the chip formation process in micro-cutting when the uncut thickness is low enough and is below the critical value of the minimum uncut chip thickness. In this case, the inflowing work material is ploughed, rubbed, and pressed at the rounded cutting edge with possible elastic recovery at the flank face. When the uncut chip thickness is much smaller than the minimum uncut chip thickness, the inflowing work material moves beneath the cutting edge and undergoes elastic deformation beneath the cutting edge and eventually recovers the elastic deformation. As the uncut chip thickness increases, elastic–plastic deformation occurs and only a certain portion of the work material deforms elastically with the portion adjacent to clearance face undergoing plastic deformation. When the uncut chip thickness crosses the minimum chip thickness level, material separation starts and shearing mechanism plays a significant role in the chip formation. In this case, ploughing and shearing mechanisms are simultaneously present where the material above the depth of the minimum uncut chip thickness forms chip and the material below the minimum uncut chip thickness level moves beneath the cutting edge, undergoes deformation under the cutting edge with a partial elastic recovery. For much higher uncut thicknesses, ploughing becomes less dominant and cutting takes place primarily in shearing dominant regime and the elastic recovery at the flank face is negligible.

Several research studies have been made to evaluate the minimum uncut chip thickness, both analytically and experimentally. These research attempts helped to infer that MUCT is dependent on the cutter-edge radius and the properties of the work material, and has profound influence on the surface generation and cutting forces. Micro-cutting of steel revealed that the sharpness of the cutting tool influences the critical cutting depth at which chip separation occurs from the material irrespective of different microstructures. Examination of the topography of the micro-cut surfaces also indicated the decisive influence of the minimum cutting depth on the surface roughness (Weule et al., 2001). The stagnation angle of the neutral point corresponding to the critical cutting depth can be derived analytically from the consideration of the energy expended in ploughing and the energy utilized in chip formation. Experiments were conducted on steel specimens using blunt tools with uncut thickness comparable to the cutting-edge radius to observe the onset of chip formation and transition from

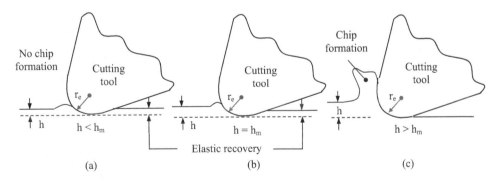

Figure 1.5 Effect of minimum chip thickness, edge radius of cutting tool, and uncut material depth on chip formation.

ploughing to cutting (Basuray et al., 1977). Ultra-precision cutting using diamond cutting tools with cutting edge of few nanometers and achieving a nominal minimum thickness of cut around 1 nm describes the minimum thickness of cut revealing extreme accuracy attainable in ultra-precision cutting. The analysis of the chip removal of this type of nanoscale ultra-precision cutting can be done with atomistic or molecular dynamics simulation as the minimum thickness of cut is in the range of few nanometers with material removal taking place in a limited region containing only a few atoms (Ikawa et al., 1992). Experimental analysis of diamond turning on Al alloys revealed that diamond cutting tool sharpness exerts a significant influence on the minimum cutting thickness and a much sharper tool was needed to reach a much smaller cutting thickness. A minimum thickness in the range of 0.05–0.2 µm could be achieved using a diamond cutting tool having edge radius in the range of 0.2–0.6 µm. Experimental results also indicated that tool sharpness influenced surface integrity of the machined surface and sharper tool resulted in lower surface roughness, lower micro-hardness, and lower residual stress of the machined surface layer (Yuan et al., 1996). Intermittency of chip formation is observed in micro-milling due to low projected feed per tooth and low engagement of the cutting edge. The cutting forces varies with tooth passes as the material gets accumulated in several non-cutting tool passes and non-periodicity of cutting forces are observed due to the dominant presence of minimum chip thickness effect. Minimum chip thickness can be estimated in full immersion micro-milling from the cutting force data under various cutting conditions. In a study of micro-milling of brass, the ratio of minimum uncut chip thickness to the cutting-edge radius was observed to lie within 0.16–0.25 (Kim et al., 2004). However, encompassing several studies in micro-cutting, the ratio of minimum uncut chip thickness to the cutting-edge radius was observed to lie within the range of 0.10–0.40. In ultra-precision micro-cutting where the cutting depth is much less than the cutter-edge radius, an expression relating minimum cutting thickness, cutting-edge radius, and average coefficient of friction of the tool–work combination can be derived considering the force equilibrium for the work material at the stagnation point corresponding to the limiting cutting depth from where it may be assumed that work material neither moves to the chip nor moves down to get ploughed at the rounded edge. Theoretical analysis expressing minimum cutting thickness as a function of cutting-edge radius, coefficient of friction for the tool–work material, and the ratio of radial to tangential forces at the point corresponding to the minimum thickness established the relationship between the sharpness of the cutting edge and the minimum cutting thickness. A study consisting of a series of micro-cutting experiments on materials, namely, aluminum, brass and oxygen-free high conductivity (OFHC) copper, estimated the minimum cutting thickness based on the aforementioned concept, and the cutting forces and machined surface integrity at different cutting depths were analyzed. Experimental results of these tests showed that near to the minimum cutting depth, the thrust (normal) force was comparably higher than the principal cutting force (tangential force) due to considerable ploughing and burnishing effect. Examining quality of machined surface revealed that the best surface condition was achieved at lowest cutting depth that was above the minimum cutting depth (Son et al., 2005). Ratio of minimum uncut chip thickness to cutting-edge radius referred as normalized minimum chip thickness can be predicted in micro-cutting based on molecular mechanical theory of friction and slip line model of micro-cutting. Minimum chip thickness can be experimentally determined using ultra-precision optical profiler by measuring the sudden surface height change on the generated sidewall surface due to the transition from ploughing and rubbing to the chip formation in micro-end milling. In a study, the normalized minimum chip thickness was found to increase as the cutting velocity and tool edge radius increased in micro-cutting carbon steels. This study further showed that the higher the carbon content in carbon steels, the larger the normalized minimum chip thickness. The normalized minimum chip thickness was found to remain uniform over a range of cutting velocities (60–500 m/min) and tool edge radii (0.5–5 µm) in micromachining Al6082-T6 work material due to thermal softening effect and strain hardening effect canceling out each other (Liu et al., 2006). The minimum thickness of cut can be evaluated experimentally by gradually engaging the cutting tool with the workpiece with increasing depths and identifying the zone that characterizes gradual transition from ploughing to cutting. This zone that includes the region where chip separation starts and corresponds to the minimum thickness of cut experiences a sharp increase of roughness parameter R_z (average maximum peak to valley height of five consecutive sampling lengths within the evaluation length) (Ramos et al., 2012).

1.2.3   Size Effect in Micro-cutting

The location of the buildup zone and the stagnation point depend on the work material, cutting tool, cutting-edge geometry, and cutting conditions. The stagnation point separates the micro-cutting zone into two regions: the shear dominant region and the ploughing and rubbing region. The region where the material flows beneath the bottom arc segment of the rounded cutting edge and undergoes severe rubbing action is the rubbing region. The effect arising due to low uncut material depth, micro-tool geometry, material characteristics, and process mechanics in micro-domain cutting influences the cutting forces, specific cutting energy, surface quality, etc. apart from efficient material removal. In micro-cutting, for removal of smaller and smaller amounts of material, the cutting force needed to cut is larger than that anticipated at lower uncut chip thicknesses. The implication of this is more pronounced in the specific cutting energy, defined as the energy needed to remove a unit volume of work material. This effect is depicted in Figure 1.6 where the dashed portion of the cutting force variation indicates anticipated values and the solid line indicates the actual values. At low uncut chip thickness, significant amount of energy is utilized in ploughing, rubbing, and plastic deformation of the machined surface without any significant material removal.

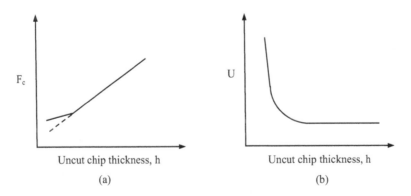

Figure 1.6 Variation of (a) cutting force (F_c) and (b) specific cutting energy (U) with uncut chip thickness (h) in micro-cutting.

These lead to the increased specific cutting energy at low feed rates and this phenomenon is described as the "size effect" (Lucca et al., 1993; Liu and Melkote, 2006; Malekian et al., 2009; Ray et al., 2020b). The size effect on the machining characteristics in micro-cutting is mainly attributed to the increased ploughing and work material strengthening at low cut size. Work material strengthening arises due to the lack of defects or preferential sites from where the shear deformation originates. In micro-cutting, low ratio of uncut chip thickness to the cutting-edge radius and the ploughing phenomenon significantly contribute to the inefficient material removal, increased cutting forces, larger specific cutting energy, vibrations and process instability, decreased surface finish and poor edge generation due to the development of burrs, etc. The process parameters need to be suitably chosen in micro-cutting operations to improve the desirable process outcomes, namely, efficient material removal, low cutting forces, better surface finish, less burr generation, etc.

1.2.4   Work Material Property and Microstructure Effect in Micro-cutting

Different types of material defects are inherently present in the workpiece depending upon the type of material. Crystalline materials contain dislocations in the lattice structure, whereas in amorphous materials, defects can be in the form of micro-voids and micro-cracks. These defects help in the material deformation process. Plastic deformation occurs within a small volume of the work material at small scales of material removal in micro-cutting. Depending upon the strain, strain rate, strain gradient, and the temperature in the cutting zone, the flow stress during deformation influences the material's resistance to the plastic deformation, machining forces, energy consumption, etc. As the number of defects is less in the microscale volume during micro-cutting, the strength of the work material is not reduced to the extent that can be expected from a scaled-down macro-cutting process. The apparent increase of material strength that influences the cutting forces and specific cutting energy gives rise to the size effect. The uncut chip thickness in micro-cutting is often comparable to the grain size in crystalline substances. Therefore, crystallographic orientation, presence of dissimilar phases, and grain-level anisotropy largely influence the cutting process. As grain sizes are much comparable to the micro-cutter-edge radius, often the cutting takes place across the grain. For coarse grain sizes, the effects of the crystallographic orientation, grain boundaries, etc. make the material response in cutting to be heterogeneous and anisotropic, whereas fine or ultrafine grains make the response of the material related to chip removal comparatively isotropic. Ultrafine grained materials may however show increased strength due to larger specific grain boundaries that causes increased resistance to the dislocation movements.

Amorphous materials like bulk metallic glasses are amorphous metallic alloys having non-directional metallic bonds without long-range atomic order and exhibit higher strength than many conventional alloys due to lack of grain boundaries or crystal defects. Amorphous materials may be considered to possess a high degree of homogeneity and can be macroscopically considered isotropic. Deformation under applied load in bulk metallic glasses (BMGs) gets highly localized into shear bands. Generally, at low temperature and low strain rate of deformation, bulk metallic glasses macroscopically exhibit brittle nature. However, in micro-cutting of bulk metallic glasses, due to deformation at high strain rate, associated temperature rise, low thermal conductivity of the work material, etc., nanocrystal particles that are formed within the amorphous matrix help to develop multiple shear bands leading to enhanced plasticity in some BMGs (Schroers and Johnson, 2004; Liu et al., 2007; Chen, 2011; Wang et al., 2004). Analysis of surface microstructure of bulk metallic glass in high-speed milling process showed that the increase in feed rate resulted in the increase of nanoscale crystalline precipitates in the amorphous matrix (Maroju et al., 2018). Shear transformation zones that develop embryonic shear bands in metallic glasses propagate rapidly at the yield point leading to the failure of the work material. Studies indicated that in small-scale deformation processes of amorphous metallic alloys, transition occurred in the material from a strong yet brittle state to a stronger and ductile state (Jang and Greer, 2010; Kuzmin et al., 2012). The deforming volume is substantially reduced in micro-cutting causing increased strengthening in bulk metallic glasses (BMGs), as the rapid propagation of shear bands is suppressed due to lack of preferential sites from where the embryonic shear band originates. For ductile materials, different types of cutting conditions induce different types of chip formation which influences the cutting process. Uncontrolled fracture and crack propagation that occur during cutting in brittle materials such as glass lead to poor surface generation and subsurface damage. These may be avoided with a much lower depth of cut below a critical thickness. Micromachining at this much reduced chip thickness with very small bits of material removal causes plastic flow in the material and is referred as ductile mode of cutting that helps to avoid uncontrolled fractured surfaces.

Mechanical and physical properties of work material such as ductility, hardness, yield strength, strain hardening, thermal conductivity, material microstructure, microstructural defects, and atomic arrangements influence the mechanism of chip formation and the micro-cutting process depending on the cutting conditions employed in micro-cutting. All these material-related factors have to be taken into consideration for the understanding of the micro-cutting processes. Equal channel angle processing (ECAP) is an effective cold working technique for grain refinement of material. Experimental analysis of the effect of material microstructure and grain size on the machined surface integrity in micro-milling indicated that surface roughness produced by micro-milling is highly dependent on the material grain size. Experimental study using aluminum workpieces having different grain sizes processed through different techniques such as hot extruded and annealed Al5083, conventionally processed cold extruded Al5083, and equal channel angle (ECAP) processed Al5083 showed that the surface roughness of the ECAP processed ultrafine grained (UFG) sample was lowest and at least three to four times lower than the other two types. The study indicated that grain refinement improved surface integrity (Popov et al., 2006). Finite element analysis of the effect of microstructure on the chip formation in micro-milling of steel having heterogeneous microstructure (multiple phases in the microstructure, with harder pearlitic phase embedded in the softer matrix of ferrite phase) indicated that the plastic strain distribution throughout the cut chip thickness was largely irregular where severe strain localization and large plastic strain occurred in the softer phase that embedded the less plastically deformed harder phase. The plastic strain pattern for heterogeneous material appeared to be discontinuous in the primary shear zone. Experimental analysis of the generated chips in normalized AISI 1045 showed ripples and dimples on the free surfaces of the chips. Presence of surface dimples and rippled chip surface in experimental chips corresponded well with the heterogeneous finite element model that captured satisfactorily the surface imperfections that actually occurred in the work samples (Simoneau et al., 2007). Investigation on the influence of grain size and material microstructure on the resulting surface quality in micro-cutting pure copper workpiece samples showed that ultrafine grained (UFG) samples with average grain size of 0.2 μm had lower surface roughness than the normalized coarse grained (CG) samples with average grain size of 30 µm. Hardness analysis of the micro-milled surface indicated that UFG samples had higher hardness than CG samples. Surface condition analysis revealed that observable surface defects in UFG samples were less than the CG samples. For the UFG samples, micro-burr formation was much less compared to the CG samples (Elkaseer et al., 2016). Analysis of cutting forces and power spectrum density in ultraprecise cutting of different materials with polycrystal, monocrystal, and amorphous microstructures under different cutting conditions showed that smaller grain size in polycrystalline material had wider frequency content compared to the larger grain-sized materials. Moreover, large dynamic cutting and thrust force variations were observed at the grain boundaries. Significant increase of specific cutting force at finer depth of cuts was observed. Study revealed that in comparison to the polycrystal materials, single-crystal and amorphous materials showed less variation of the dynamic forces due to homogeneous and isotropic property of the materials. A study indicated that better cutting performance was obtained in materials having ultrafine grains and uniform microstructure with depth of cuts much larger than the grain sizes (Furukawa and Moronuki, 1988).

1.2.5   Chip Formation in Micro-cutting

Analysis of chip formation and study of chip morphology reveal the underlying deformation and fracture mechanisms prevailing in machining processes. Understanding the fundamental mechanisms of chip formation helps in understanding the influence of the process variables on the micro-cutting process and is critical to optimize the machining performance and development of improved micro-cutting tools and processes. Investigations reported by several researchers indicate that workpiece material property, cutting conditions, cutting tool material, and geometry have significant influence on the plastic deformation and chip formation process. In a study involving micro-milling of different workpiece materials, for example, copper (OFHC), Al 6082-T6, AISI 1005 steel, AISI 1045 steel, Ti-6Al-4V, Inconel 718 carried out at different feed rates (feed per tooth: 0.02–10 µm) revealed the characteristics of the chip morphology of the materials at different cutting conditions, namely, when the uncut chip thickness was below and above the cutter-edge radius. Smaller segmented and discontinuous chips were produced for all the materials at lower feed rates. Micromilling of the Ti alloy at lower feed rates produced chips that were longer in length but smaller in thickness compared to other materials. Analysis of chip morphology with uncut chip thickness above the cutter-edge radius revealed that the materials produced segmented chips excluding copper which produced discontinuous chips. AISI 1005 steel chips showed higher segmentation than AISI 1045 steel. Nano-hardness measurements showed that chips had higher hardness compared to the bulk material, indicating full permeation of plastic deformation in the chip material. The analysis of the acoustic emission (AE) energy distribution of the materials in different frequency bands represented different deformation and fracture mechanisms involved in chip formation process (Mian et al., 2011). Chip morphology study in micro-turning of medium carbon steel revealed that as the uncut chip thickness was reduced and brought below the average grain size, the continuous form of chip changed to quasi-shear-extrusion chip with alternate layers of hard pearlite and soft ferrite where the softer ferrite was extruded between the hard pearlitic grains creating large plume of ferrite at the chip free surface. The free surface of the chips revealed lamellar structure with distinct slip-line along the shear front. The under surface of the chips at the tool–work interface showed signs of severe plastic deformation and significant rise of temperature. Mechanics of formation of quasi-shear-extrusion chip in microscale cutting process across alternating grains of hard and soft material in steel samples can be analyzed using finite element model (Simoneau et al., 2006). A study on chip formation in micro-cutting of cold-rolled brass revealed that serrated chips with regular shear band spacing were formed for different uncut chip thicknesses (10–20 μm) when using a cutting tool having 0° rake angle. However, using tool with negative rake angle (–25°) resulted in chips having irregularly spaced coarse shear bands identified in the primary deformation zone (Wang et al., 2010). Chip morphology study in micro-end milling of aluminum 6082-T6 using diamond cutting tool in the feed rate range 0.2–5 μm and different cutting speeds showed that the upper side of the chips had less smooth surface than the lower side surface. In this study, the chip thickness was found to increase with increased stress at the tool–work interface on the rake face and rougher lower surface of the chips as the feed rate was increased. However, cutting with tungsten carbide tools resulted in longer chips than that obtained with diamond tools under similar cutting conditions. In cutting with tungsten carbide tools, the chip thickness ratio was observed to be much smaller than that obtained with diamond tools and the values increased with the increase in feed rate. Furthermore, the study indicated no appreciable influence of the cutting speed on the chip morphology (Niu et al., 2018). Study of chip morphology at different feed rates in micro-turning of pure copper and annealed pure copper of different grain sizes revealed that the decrease of the feed rate changed the chip from continuous to segmented form in both original (un-annealed) and annealed pure copper. The annealed copper had higher grain sizes and lower hardness than the pure copper. The chips generated in machining of annealed pure copper were more segmented than in the original copper. Analysis of the chip morphology further revealed that the free surface had lamellar pattern with comparatively rougher surface than the back surface. The lamellar pattern of chips might be due to the dislocation gliding from the inside to the free surface along the gliding plane (Yu et al., 2015). Investigation on the chip formation in micro-cutting of Fe-based amorphous alloy showed that the chips were having lamellar structure due to the periodic formation of the localized shear bands and adiabatic deformation in the primary deformation zone (Ueda and Manabe, 1992). Analysis of the chip segmentation in cutting Zr-based bulk metallic glass, an amorphous alloy, revealed that the cutting zone temperature and the free volume influenced shear localization, shear flow instability, frequency of chip segmentation, etc. (Maroju and Jin, 2019).

1.2.6   Material Removal Mechanism in Micro-milling Process

Chip formation in micro-cutting is a highly dynamic process. The dynamic chip thickness is highly non-uniform due to the process dynamics, tool run-out, tool wear, vibrations, regenerative effects, etc. Chip formation in micro-cutting experiences changes in the shear angle of the slip planes in the cutting zone and variation in the chip thickness. Depending upon the material properties and the process conditions, wavy, segmented, discontinuous, elemental, and particle chips are formed in micro-cutting. Micro-channels are common micro-features that are often produced by full immersion micro-milling process where the micro-cutter diameter determines the width of the micro-channel and the axial depth of cut determines the micro-channel depth. In orthogonal shaping or turning processes, the uncut chip thickness may be considered uniform provided the vibrational and regenerative effects are not considered. However, in milling process, the uncut chip thickness is not uniform. If a full immersion micro-end milling process geometry is observed as shown in Figure 1.7, it is

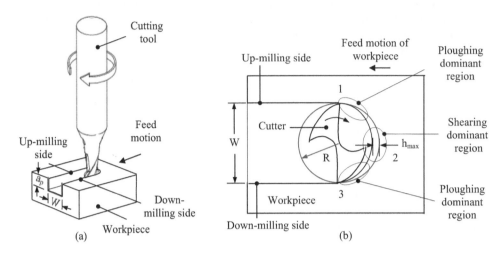

Figure 1.7 (a) Full immersion micro-end milling. (b) Geometry of micro-channel (slot) cutting in end milling process by two fluted micro-end mill of radius *R*; width of the micro-channel *W*.

found that the instantaneous uncut chip thickness is not constant; rather, it varies with the rotation of the cutting tool and the immersion angle of the cutting point. The instantaneous uncut chip thickness is zero when a cutting edge starts cutting during its initial engagement. Gradually, it increases to a maximum value nearly equal to the feed per tooth at the center of the micro-channel and again becomes zero at the exit from the cut. A chip is not developed until the instantaneous uncut chip thickness exceeds the minimum uncut chip thickness and the elastic recovery of the uncut layer of the work material occurs. When the instantaneous uncut chip thickness exceeds the minimum uncut chip thickness, a chip starts to form. Due to the small ratio of uncut chip thickness to cutting-edge radius, considerable ploughing may still exist at this portion. Again, at the exit side, the chip generation stops when the instantaneous uncut chip thickness becomes less than the minimum uncut chip thickness. At the entry of the cutting edge into the cut, the region from the entry point to the point where clear chip separation starts to occur (region 1), and at the exit of the cutting edge from the cut, the region from the point where clear chip separation seizes to occur to the exit point (region 3), are predominantly ploughing dominant regions. The middle of the micro-channel passage (region 2) where chip separation takes place clearly (provided feed per tooth value is more than MUCT) is the shearing dominant region as the chip formation process is mainly due to shearing. One side of the micro-channel experiences up-milling cut, while the other side experiences down-milling cut.

For a two fluted cutter in full immersion milling, when a cutting edge completes its excursion in the cut, the subsequent cutting edge encounters the material left from the earlier cutting teeth pass. For the region in which a chip was not generated with the previous cutting edge, the current cutting edge experiences a larger chip thickness. Thus, the initial point of chip formation moves back to an earlier angular position (smaller immersion angle than the earlier one), and the actual material cutting region is extended. When the projected feed per tooth at a position is less than the minimum uncut chip thickness, chips are generated once in several rotational passes till a sufficient material thickness accumulates ahead of the cutting

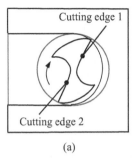
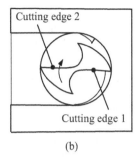
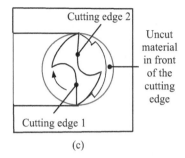

(a)   (b)   (c)

Figure 1.8 Micro-milling with minimum chip thickness effect. (a)–(c) Work material left uncut by cutting edge 1 and excess material accumulates ahead of cutting edge 2 resulting in the increase of the ratio of measured chip volume to nominal chip volume.

edge (Filiz et al., 2007). The analysis of chip volumes in a study of micro-end milling of brass at different feed rates applicable in micro-cutting revealed that the ratio of measured chip volume to nominal chip volume approached the value of unity at higher feed rates, whereas, at lower feed rates, the ratio was much higher indicating that chips were not formed in each pass of a cutting tooth. In this study, examination of feed marks revealed that lower feed rates resulted in several non-cutting passes (Kim et al., 2002). These aspects have been illustrated in Figure 1.8. Added to this effect of several non-cutting passes, the tool run-out effect that arises due to improper tool symmetry, imperfect tool alignment in tool holder, mismatch between tool holder and spindle, positional error in spindle bearings, etc. resulted in the variation of the chip load on the cutting edges leading to significant force variation, rougher surface generation, unequal tool edge wear, etc.

1.3   MICRO-MILLING PROCESS CHARACTERISTICS

1.3.1   Micro-milling Forces and Specific Cutting Energy

Cutting forces are important process responses in mechanical micro-cutting processes. Conventional cutting force model with sharp-edged cutting tool cannot be adopted in micromachining as the uncut material thickness is most often comparable to the edge radius of the cutting tool and the effects of shearing and ploughing are to be considered to analyze the cutting mechanics. Cutting forces are related to the material removal mechanism and chip formation. Non-uniformity in material microstructure can cause cutting force fluctuations. Experimental analysis related to micro-cutting indicated that smaller material grain size resulted in larger cutting forces and higher specific cutting energy. Studies however indicated that influence of the cutting-edge radius was larger than the influence of the material grain size on the cutting forces and the specific cutting energy (Wu et al., 2016). Improper cutting conditions may lead to excessive cutting force generation, excessive tool deflection and bending, etc., causing tool failure. Gradual tool wear causes increase in the cutting-edge radius resulting in higher cutting forces. The magnitude of the cutting forces in micro-milling is smaller as compared to macro-milling processes. The force amplitudes are generally observed to be around few tenths to few tens of Newtons. However, the specific cutting forces in micro-milling are comparable to the macro-milling process as uncut chip thickness is reduced (Newby et al., 2007). Cutting force signatures in micro-milling contain fluctuations and shows erratic behavior particularly at low feed rates when ploughing effect is prevalent and clear chip removal is not taking place (Filiz et al., 2007; Ray et al., 2020c). Figure 1.9 depicts the comparison of predicted and experimental cutting force signatures at two different feed rates in micro-milling bulk metallic glass. The cutting tools used in the experiments were tungsten carbide flat-end micro-end mills. It is evident that at low feed rate, actual cutting force profile largely deviated from the predicted values in comparison to cutting when carried out at high feed rates. Various factors such as vibration, tool run-out, and tool wear that influence process dynamics are responsible for the erratic behavior. The quality of the micromachined surfaces is influenced by the static and dynamic characteristics of the cutting forces. Modeling micro-cutting forces is useful in characterizing the micro-cutting mechanics.

Mechanistic modeling in micro-milling using unified mechanics of cutting has been followed by many researchers. In this approach, the cutting force coefficients were obtained using oblique cutting analysis and orthogonal cutting tests data, taking suitable material model into consideration (Srinivasa and Shanmugam, 2013). In some analyses, the interference volume of the workpiece with the rounded edge of the cutting tool were considered to model the ploughing phenomenon (Malekian et al., 2009; Lu et al., 2015; Jun et al., 2012).

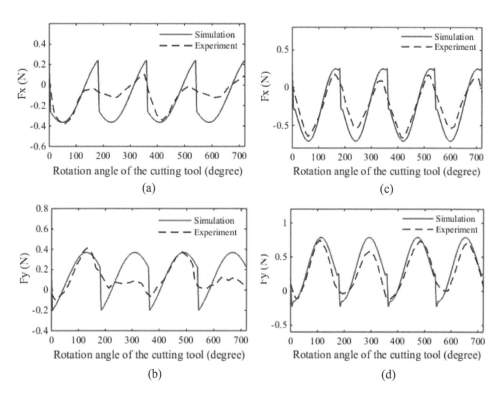

Figure 1.9 Predicted and experimental cutting forces (a)–(b) at feed rate 0.5 µm/tooth, cutting velocity 0.062 m/s, axial depth of cut 20 µm; (c)-(d) at feed rate 6 µm/tooth, cutting velocity 0.062 m/s, axial depth of cut 20 µm (Ray, 2020).

Micro-milling cutting force model also used separate edge coefficients for the ploughing and shearing dominant regime computed from average cutting forces where the ploughing forces are modeled as proportional to the interactive area between the rounded cutting edge and the work material (Ray et al., 2020c). According to this model in the ploughing-dominant regime of cutting ($h < h_m$), the tangential, radial, and axial cutting forces acting on a elemental flute element of the flat-end micro-end mill of width dz at a height of z from the tip of the end mill and denoted by $dF_{tp}(\theta, z)$, $dF_{rp}(\theta, z)$, $dF_{ap}(\theta, z)$, respectively, are modelled as follows:

$$
\left.
\begin{aligned}
dF_{tp}(\theta,z) &= \left[K_{tp}\, l_p(\theta,z) + K_{tel} \right] dz \\
dF_{rp}(\theta,z) &= \left[K_{rp}\, l_p(\theta,z) + K_{rel} \right] dz \\
dF_{ap}(\theta,z) &= \left[K_{ap}\, l_p(\theta,z) + K_{ael} \right] dz
\end{aligned}
\right\} \qquad 1.1
$$

Here $\theta(z)$ is the radial immersion angle of the cutting point; K_{tp}, K_{rp}, K_{ap} are the ploughing coefficients in tangential, radial, and axial directions, respectively, and represent the intensity of ploughing and rubbing. K_{tel}, K_{rel}, K_{ael} are the edge force coefficients in tangential, radial, and axial directions, respectively, and represent the intensity of edge forces and incorporate the frictional effects in the ploughing dominant regime of cutting. As presented in Equation (1.1), l_p represents the interactive length of the inflowing work material with the bottom segment of the rounded cutting edge and a small portion of the flank face and $l_p dz$ represents the elemental interactive area of the inflowing work material of width dz subjected to ploughing. The schematic of the cutting zone of the micro-model is depicted in Figure 1.10. Due to the curved edge, the ploughing forces act along the radius of the cutting edge (r_e) and the sliding forces act on the flank face due to the elastic recovery (Δ_e) and interaction of the work material at the flank face. As shown in Figure 1.10, O is the center corresponding to the edge radius of the cutting tool. Point Q is the point on the cutting edge corresponding to the limiting uncut chip thickness (h_m). Material above point Q flows as chip, whereas material below point Q is ploughed below the cutting edge and moves beneath the flank face. As shown in Figure 1.10, $\angle ECF = \alpha_n$ (clearance angle); $\angle QOB = \alpha_p$ (ploughing angle).

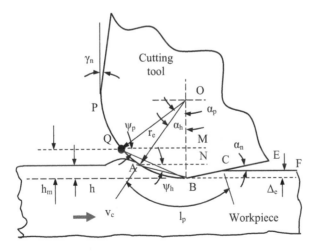

Figure 1.10 Exaggerated view of the cutting zone when $h < h_m$.

In the shearing dominant regime of cutting ($h > h_m$), the elemental cutting forces $dF_{tc}(\theta, z)$, $dF_{rc}(\theta, z)$, $dF_{ac}(\theta, z)$ are modeled as follows:

$$\begin{aligned}
dF_{tc}(\theta, z) &= \left[K_{tc} h(\theta, z) + K_{te2} \right] dz \\
dF_{rc}(\theta, z) &= \left[K_{rc} h(\theta, z) + K_{re2} \right] dz \\
dF_{ac}(\theta, z) &= \left[K_{ac} h(\theta, z) + K_{ae2} \right] dz
\end{aligned} \right\} \qquad 1.2$$

Here $\theta(z)$ is the radial immersion angle of the cutting point, $h(\theta, z)$ represents the uncut chip thickness at the cutting point. K_{tc}, K_{rc}, K_{ac} are the cutting (shearing) coefficients in tangential, radial, and axial directions, respectively, representing shearing of the work material and K_{te2}, K_{re2}, K_{ae2} are the edge force coefficients in tangential, radial, and axial directions, respectively, representing frictional effects at the cutting edge in shearing dominant regime. From Equations (1.1) and (1.2), the expressions for average forces per tooth period are evaluated for the ploughing and shearing dominant regions in selected X, Y, and Z coordinate directions. A set of micro-milling experiments was done such that the entire cutting takes place in the ploughing dominant regime where feed per tooth is less than minimum uncut chip thickness ($f_t < h_m$) and also at different feed rates such that feed per tooth is greater than minimum uncut chip thickness ($f_t > h_m$). The experimental data were utilized to evaluate the specific cutting force coefficients. Based on the proposed model and the estimated coefficients, cutting force simulations in micro-slot end milling were done using a computational program. Slip Line Field model is another widely used method to calculate micro-cutting forces (Waldorf et al., 1998; Jin and Altintas, 2011). Numerical methods using finite element (FE) analysis are now popularly used for prediction of micro-cutting forces. Cutting force coefficients evaluated from FE simulations were used in the mechanistic formulation of the cutting forces. The cutting force coefficients were estimated from a series of FE simulations for a wide range of chip loads and edge radii (Jin and Altintas, 2012). Micro-milling cutting forces can be deduced in assisted micro-milling process as reported in literatures (Melkote et al., 2009; Kumar and Melkote, 2012), where a laser beam is focused on the work material ahead of the cutting path to soften the material or reduce the flow stress in work material.

Specific cutting energy is a measure of the energy expended to remove unit amount of material. It is an effective machining parameter by which the efficiency of a metal cutting process or the machinability of a work material for a particular process can be estimated from cutting power and material removal rate. The size effect in micro-cutting is typically characterized with the abrupt non-linear increase of specific cutting energy with decrease of uncut chip thickness at values below edge radius of the cutting tool. Several research studies have been done to understand the size effect in micro-cutting (Ray et al., 2020b; Zhang et al., 2017; Ng et al., 2006; Zhong et al., 2016; Lauro et al., 2015). Studies showed that the size effect in micro-cutting is mainly influenced by ploughing, and work material microstructure and strengthening at low cut size. It is interesting to note that the size effect on the machining characteristics in micro-milling amorphous metallic alloys such as bulk metallic glass may also be attributed to the increased strengthening of the work material as in crystalline materials. However, the micro-mechanism behind the strengthening effect is different from that observed in crystalline counterpart, which is already highlighted in Section 1.2.4. Figure 1.11 depicts the size effect in micro-milling bulk metallic glass. As shown in Figure 1.11, the predicted values presented are estimated following a micro-cutting model. The predicted results are compared with the experimental results carried out at two different cutting speeds u_{s1} (0.062 m/s) and u_{s2}

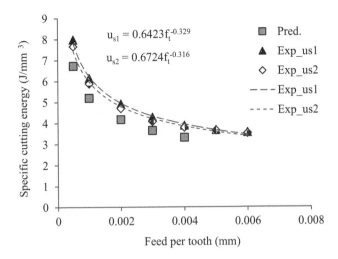

Figure 1.11 Analytical and experimental measures of specific cutting energy (Ray, 2020).

(0.094 m/s), respectively. The micro-cutting zone assumed in the model is depicted in Figure 1.12, where work material layer with uncut thickness h approaches the cutting edge and gets bifurcated from the tip of the buildup zone ABC to either form chip or pushed below the bottom curved segment of the rounded cutting edge into the machined surface. The predictive specific cutting energy (U_c) is estimated as follows:

$$U_c = U_{cc} + U_{cr} \frac{r_e \left[1 - \cos \alpha_s \right]}{h} ; \quad U_{cc} = \frac{F_{cc} \cdot 1}{h \cdot 1 \cdot 1} = \frac{F_{cc}}{h} ; \quad F_{cc} = F_{cc1} + F_{cc2}$$

$$F_{cc1} = \left[p_1 \cos \gamma_e \frac{\left[h - r_e (1 - \cos \alpha_s) \right]}{\cos \gamma_e} - k \sin \gamma_e \frac{\left[h - r_e (1 - \cos \alpha_s) \right]}{\cos \gamma_e} \right]$$

$$F_{cc2} = k \left[r_e \cos \gamma_e - \tan \gamma_e \left\{ h - r_e (1 - \cos \alpha_s) \right\} - r_e \sin \alpha_s \right]$$

$$U_{cr} = \frac{k \left[\pi \sin \alpha_s \tan \alpha_s + \dfrac{2\alpha_s}{\cos \alpha_s} \right]}{\left[1 - \cos \alpha_s \right]} \tag{1.3}$$

Here F_{cc1} and F_{cc2} are the forces acting on the face AB and BC per unit width of cut in the horizontal direction and expressed in terms of pressure p_1 and shear yield stress k (Figure 1.12).

1.3.2 Micro-milling Quality Characteristics

1.3.2.1 Surface Roughness

Surface roughness is a common indicator to assess the quality of the generated surface. Every micromachining process aims to create functional surfaces that perform the intended service

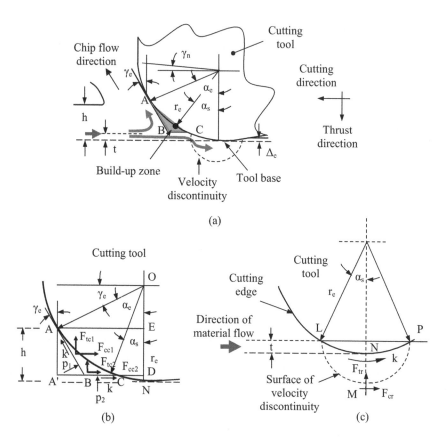

Figure 1.12 Schematic of the micro-cutting zone and representation of the cutting forces. (a) Work material flow around rounded cutting edge. (b) Forces in the cutting region. (c) Forces in the rubbing region below lower curved surface of the cutting edge (Ray, 2020).

properly and meet the quality requirements. Researchers have carried out several studies to explore different aspects like surface generation, dimensional accuracy, surface roughness related to micromachining applied on various types of materials. Quality of micro-components that are made using precision manufacturing technology depends strongly on the surface and edge quality. Process kinematics in mechanical micro-cutting, including micro-milling, influences surface quality. For a tool–work combination in micro-milling, the mode of milling, for example, up-milling or down-milling, can reveal variation in the surface roughness. Cutting tool edge geometry influences the extent of surface roughness generation for different levels of feed rate. Depending on minimum uncut chip thickness, very low feed rate causes significant ploughing. Excessive ploughing leading to work material plastic side flow and elastic recovery causes non-uniform rise of surface ridges and increase in surface roughness. Research studies explored the influence of cutting conditions and the effect of work material microstructure on the generated surface roughness (Bissacco et al., 2005; Vazquez et al., 2010; Kiswanto et al., 2014). Studies indicated that in micro-cutting engineering materials, surface roughness increases with the increase in feed rate when uncut chip thickness is more than the edge radius of the cutting tool. Decrease in feed rate shows an increase in the surface roughness for uncut chip thickness lower than edge radius of the cutting tool. For a range of feed rates, surface roughness decreases with decrease in feed rate, reaches a minimum, and then tends to increase with further reduction in feed rate. This phenomenon may be identified as a "size effect" of surface roughness. At very low feed rates in micro-milling, some of the plastically deformed material that is ploughed aside causes a burr-type formation on the feed mark ridges and sticks on the newly machined surface (Ray et al., 2020a). This deteriorates the machined surface quality and increases the generated surface roughness. Figure 1.13 depicts the response surface plots for average line roughness (R_a) in full immersion micro-end milling of Zr-based bulk metallic glass using carbide flat-end micro-end mills. Furthermore, multi-phase work material developed higher surface roughness than the single-phase material (Vogler et al., 2004).

Process dynamics, machining vibrations, process stability, motion errors of different linear and rotary drives greatly influence the achievable surface quality (Kline et al., 1982). Apart from these, several other stochastic factors, for example, abrupt loss of edge geometry and cutting-edge serrations, influence the surface roughness. Characterization of the micromachined surface under different cutting conditions and optimization of the surface characteristics to obtain the optimal levels of cutting parameters may be a useful option to achieve desired surface quality.

1.3.2.2 Burr Formation

Burrs on the edges of micro-components have a detrimental role in their performance. Burrs are undesirable material protruding from the edges formed at different locations of the machined edges. Micro-end milling being a micro-cutting process also develop burrs having sizes comparable to feature sizes and are difficult to remove due to miniature shape of the machined features. Burr minimization and control need proper understanding of the mechanism of burr formation, and analysis of the impact of the machining parameters on the cutting process and development of burrs. Researchers have undertaken efforts to

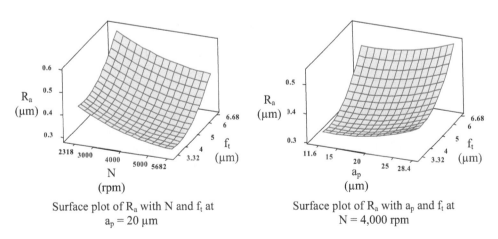

Surface plot of R_a with N and f_t at $a_p = 20~\mu m$

Surface plot of R_a with a_p and f_t at N = 4,000 rpm

Figure 1.13 Response surface plots for R_a in full immersion micro-end milling of Zr-based bulk metallic glass using 0.3-mm diameter uncoated carbide micro-end mill (Ray, 2020).

understand different aspects of micro-cutting, chip formation and burr formation in macro- and microscale machining of different materials (Gillespie and Blotter, 1976; Nakayama and Arai, 1987; Ko and Dornfield, 1991; Olvera and Barrow, 1996; Hashimura et al., 1999; Aurich et al., 2009). Modeling studies analyzed the mechanism and characterization of different types of burrs formed in micro-milling (Lekkala et al., 2011; Ray et al., 2019b). A suitable model of burr formation helps to estimate the extent of burr formation and quantify the dimension of the developed burrs. Figure 1.14 depicts the location of different types of burrs formed in full immersion micro-milling. The location of top burrs in micro-channels is such that they are difficult to remove without affecting the edge profile. An analytical model developed to estimate the width of top burrs formed in micro-end milling (Ray et al., 2019b) is expressed as follows:

$$
W_b = \frac{4}{5.97} a_p \left[\frac{(1+v)}{\sqrt{3E}} \sigma_0 e^{-\sqrt{3}\phi_a} \left[\frac{\cos\left(\phi_a + \frac{\pi}{6}\right)}{\cos\left(\phi_a - \frac{\pi}{6}\right)} - 1 \right] \right]
\tag{1.4}
$$

The parameter ϕ_a in the micro-burr formation model is estimated as follows:

$$
\phi_a = \frac{\pi}{6} - \sin^{-1}\left(\frac{\sqrt{3}p}{2\sigma_0}\right)
\tag{1.5}
$$

Here p is the radial pressure exerted by the cutting edge on the sidewall and σ_0 is the yield stress of the work material.

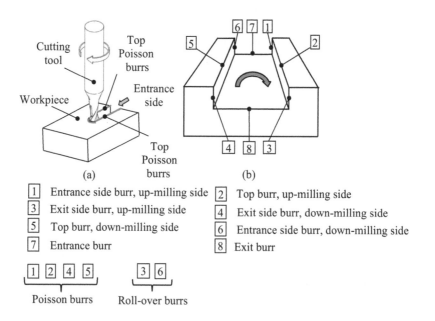

1	Entrance side burr, up-milling side	2	Top burr, up-milling side
3	Exit side burr, up-milling side	4	Exit side burr, down-milling side
5	Top burr, down-milling side	6	Entrance side burr, down-milling side
7	Entrance burr	8	Exit burr

1 2 4 5   Poisson burrs

3 6   Roll-over burrs

Figure 1.14 (a) Micro-slot milling process. (b) Location of different burrs in a micro-channel (Ray, 2020).

Figure 1.15 depicts the schematic of top burr formation mechanism in full immersion micro-milling. Study on the formation of top burrs in micro-milling indicated that axial depth of cut has significant influence on the development of burrs. The width of top burrs increased with the increase in axial depth of cut. Top burr width also increased with the decrease in feed per tooth. Analysis on the influence of feed per tooth on the width of top burrs showed that for feed per tooth less than the cutter-edge radius, burr size increases with the decrease of feed per tooth and for feed per tooth more than the cutter-edge radius, burr size decreases with the decrease in feed per tooth (Ray et al., 2019b). Figure 1.16 depicts enlarged view of the burrs in micro-channel at cutting tool entry and exit in micro-milling bulk metallic glass workpiece.

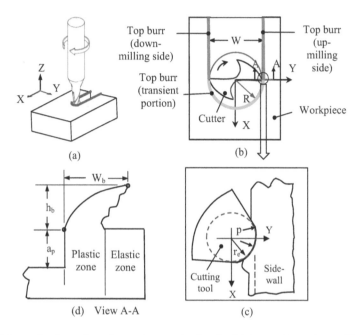

Figure 1.15 Schematic of top burr formation mechanism in full immersion micro-milling: W is the width of micro-channel; W_b is the width of top burr; and h_b is the height of top burr.

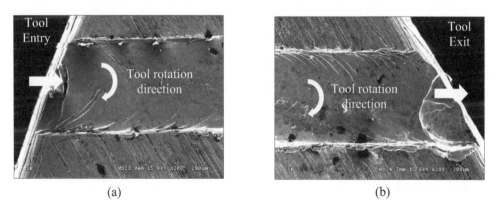

Figure 1.16 Enlarged views of the burrs in micro-channel at cutting tool (a) entry and (b) exit (Ray, 2020).

Micro-cutting performance is characterized by several response measures, namely, surface quality, part accuracy, burr formation, cutting forces, cutting energy, etc. However, micro-cutting, being an emerging technology with limited knowledge in process mechanisms and limited experience in productive manufacturing, confronts the challenge of effective industrial utilization. In micro-milling process, low stiffness and high fragility of the micro-milling tool impose limitations on the operating values of the process variables such as feed rate and depth of cut. To avoid breakage of micro-tools and excessive chip load on the cutting edges, the operating values of feed rate and depth of cut are often kept conservative. The feed rate needs to be chosen in a way that ensures efficient chip removal and also required quality of the component being machined. The depth of cut and spindle speed need to be operated at levels with a view to achieve process stability, better surface, and edge quality and lower cutting forces. Appropriate application of cutting fluid may improve surface finish, reduce tool wear, etc. However, dry cutting condition is employed to obtain cleaner production and avoid use and safe disposal of the cutting fluid having environmental implications. For effective utilization of micro-cutting in industrial practices that requires a balance between accuracy, performance, productivity, and cost-effectiveness of the process, a holistic approach with regard to optimizing work material machinability, machining quality and accuracy, material removal rate, tool life, etc. need to be addressed. In industrial process planning, selection of appropriate cutting parameters through multiple response optimization techniques may be a useful option to adopt appropriate machining strategy in achieving optimal cutting performance and improved productivity.

1.4 SUMMARY

The difference between microscale and macroscale machining primarily lies in the cutting mechanism. For both the cases, material is removed mechanically using a cutting tool having defined cutting edges. Micro-cutting shows different characteristics from macro-cutting due to the size reduction of features, tool geometry, and processing parameters. Fundamental understanding of micro-cutting mechanism started with the investigative research mainly in micro-turning and ultra-precision machining using sharp tools where uncut material thickness was kept small to achieve good surface finish and precise dimensional accuracy similar to the requirements of micro-cutting operations. In micro-cutting, the machining parameters such as feed rate is often comparable to the edge radius of the cutting tool and the grain size of the work material and the depths of cut are kept relatively smaller than the roundness of the cutting edge to prevent breakage of the cutting tool. Several critical issues related to micro-cutting and micro-milling have been discussed in this chapter and several research efforts on these critical areas related to precision machining and micro-cutting have been mentioned briefly. The aforementioned discussion may help to understand the important aspects of the interaction of the tool–work material and the machinability issues in micro-cutting.

REFERENCES

Abdelmoneim, M.E.S. (1980) Tool edge roundness in finish machining at high cutting speeds, *Wear*, 58 (1): 173–192.

Albrecht, P. (1960) New developments in the theory of the metal-cutting process part I: The ploughing process in metal cutting, *J. Eng. Ind.*, 82 (4): 348–357.

Alting, L.; Kimura, F.; Hansen, H.N.; Bissacco, G. (2003) Micro engineering, *CIRP Ann.*, 52 (2): 635–657.

Aramcharoen, A.; Mativenga, P.T.; Yang, S.; Cooke, K.E.; Teer, D.G. (2008) Evaluation and selection of hard coatings for micro milling of hardened tool steel, *Int. J. Mach. Tools Manuf.*, 48 (14): 1578–1584.

Aurich, J.C.; Dornfield, D.; Arrazola, P.J.; Franke, V.; Leitz, L.; Min, S. (2009) Burrs–analysis, control and removal, *CIRP Ann.*, 58 (2): 519–542.

Basuray, P.K.; Misra, B.K.; Lal, G.K. (1977) Transition from ploughing to cutting during machining with blunt tools, *Wear*, 43 (3): 341–349.

Bissacco, G.; Hansen, H.N.; Chiffre, L.D. (2005) Micromilling of hardened tool steel for mould making applications, *J. Mater. Process. Technol.*, 167: 201–207.

Chae, J.; Park, S.S.; Freiheit, T. (2006) Investigation of micro-cutting operations, *Int. J. Mach. Tools Manuf.*, 46 (3–4): 313–332.

Chen, M. (2011) A brief overview of bulk metallic glasses, *NPG Asia Mater.*, 3 (9): 82–90.

Connolly, R.; Rubenstein, C. (1968) The Mechanics of chip formation in orthogonal cutting, *Int. Jour. Mach. Tool Des. Res.*, 8 (3): 159–187.

Dornfield, D.; Min, S.; Takeuchi, Y. (2006) Recent advances in mechanical micromachining, *CIRP Ann.*, 55 (2): 745–768.

Elkaseer, A.M.; Dimov, S.S.; Pham, D.T.; Popov, K.B.; Olejnik, L.; Rosochowski, A. (2016) Material microstructure effects in micro-end milling of Cu99.9E, *Proc. I MechE Part B: J. Eng. Manuf.*, 1–13.

Filiz, S.; Conley, M.; Wasserman, M.W.; Ozdoganlar, O.B. (2007) An experimental investigation of micro-machinability of copper 101 using tungsten carbide micro-end mills, *Int. J. Mach. Tools Manuf.*, 47 (7–8): 1088–1100.

Furukawa, Y.; Moronuki, N. (1988) Effect of material properties on ultra precise cutting processes, *CIRP Ann.*, 37: 113–116.

Gillespie, L.K.; Blotter, P.T. (1976) The formation and properties of Machining burrs, *J. Eng. Ind.*, 98 (1): 66–74.

Hashimura, M.; Chang, Y.P.; Dornfield, D. (1999) Analysis of burr formation mechanism in orthogonal cutting, *J. Manuf. Sci. Eng.*, 121 (1): 1–7.

Huo, D.; Cheng, K. (2013) *Micro-cutting: Fundamentals and Applications*, John Wiley & Sons Ltd., 1–16.

Ikawa, N.; Shimada, S.; Tanaka, H. (1992) Minimum thickness of cut in micromachining, *Nanotechnology*, 3 (1): 6–9.

Inamura, T.; Takezawa, N.; Kumaki, Y. (1993) Mechanics and energy dissipation in nanoscale cutting, *CIRP Ann.*, 42 (1): 79–82.

Jang, D.; Greer, J.L. (2010) Transition from a strong-yet-brittle to a stronger-and-ductile state by size reduction of metallic glass, *Nat. Mater.*, 9: 215–219.

Jin, X.; Altintas, Y. (2011) Slip-line field model of micro-cutting process with round tool edge effect, *J. Mater. Process. Technol.*, 211 (3): 339–355.

Jin, X.; Altintas, Y. (2012) Prediction of micro-milling forces with finite element method, *J. Mater. Process. Technol.*, 212 (3): 542–552.

Jun, M.B.G.; Goo, C.; Malekian, M.; Park, S. (2012) A new mechanistic approach for micro end milling force modelling, *J. Manuf. Sci. Eng.*, 134 (1): 011006/1–9.

Kim, C.J.; Bono, M.; Ni, J. (2002) Experimental analysis of chip formation in micro milling, *Conf. Proc. NAMRC*, MR02–159: 1–8.

Kim, C.J.; Mayor, J.R.; Ni, J. (2004) A static model of chip formation in microscale milling, *J. Manuf. Sci. Eng.*, 126: 710–718.

Kim, J.D.; Kim, D.S. (1995) Theoretical analysis of micro-cutting characteristics in ultra-precision machining, *J. Mater. Process. Technol.*, 49 (3–4): 387–398.

Kiswanto, G.; Zariatin, D.L.; Ko, T.J. (2014) The effect of spindle speed, feed rate and machining time to the surface roughness and burr formation of Aluminium Alloy 1100 in micro-milling operation, *J. Manuf. Processes*, 16 (4): 435–450.

Kline, W.A.; DeVor, R.E.; Shareef, I.A. (1982) The prediction of surface accuracy in end milling, *J. Eng. Ind.*, 104 (3): 272–278.

Ko, S.L.; Dornfield, D.A. (1991) A study on burr formation mechanism, *J. Eng. Mater. Technol.*, 113 (1): 75–87.

Komanduri, R.; Chandrasekaran, N.; Raff, L.M. (1998) Effect of tool geometry in nanometric cutting: A molecular dynamics simulation approach, *Wear*, 219 (1): 84–97.

Kumar, M.; Melkote, S.N. (2012) Process capability study of laser assisted micro milling of a hard-to-machine material, *J. Manuf. Process.*, 14: 41–51.

Kussul, E.M.; Rachkovskij, D.A.; Baidyk, T.N.; Talayev, S.A. (1996) Micromechanical engineering: A basis for the low-cost manufacturing of mechanical micro devices using micro equipment, *J. Micromech. Microeng.*, 6: 410–425.

Kuzmin, O.V.; Pei, Y.T.; Chen, C.Q.; Hosson, J.T.M.D. (2012) Intrinsic and extrinsic size effects in the deformation of metallic glass nanopillars, *Acta. Mater.*, 60 (3): 889–898.

Lauro, C.H.; Brandao, L.C.; Carou, D.; Davim, J.P. (2015) Specific cutting energy employed to study the influence of the grain size in the micro-milling of the hardened AISI H13 steel, *Int. J. Adv. Manuf. Technol.*, 81: 1591–1599.

Lekkala, R.; Bajpai, V.; Singh, R.K.; Joshi, S.S. (2011) Characterisation and modelling of burr formation in micro-end milling, *Precis. Eng.*, 35 (4): 625–637.

Liu, K.; Melkote, S.N. (2006) Material strengthening mechanisms and their contribution to size effect in micro-cutting, *J. Manuf. Sci. Eng.*, 128 (3): 730–738.

Liu, X.; Devor, R.E.; Kapoor, S.G. (2006) An analytical model for the prediction of minimum chip thickness in micromachining, *J. Manuf. Sci. Eng.*, 128 (2): 474–481.

Liu, Y.H.; Wang, G.; Wang, R.J.; Zhao, D.Q.; Pan, M.X.; Wang, W.H. (2007) Super plastic bulk metallic glasses at room temperature, *Science*, 315 (5817): 1385–1388.

Lu, X.; Jia, Z.; Wang, X.; Li, G.; Ren, Z. (2015) Three-dimensional dynamic cutting forces prediction model during micro-milling nickel-based superalloy, *Int. J. Adv. Manuf. Technol.*, 81: 2067–2086.

Lucca, D.A.; Seo, Y.W. (1993) Effect of tool edge geometry on energy dissipation in ultraprecision machining, *CIRP Ann.*, 42 (1): 83–86.

Lucca, D.A.; Seo, Y.W.; Komanduri, R. (1993) Effect of tool edge geometry on energy dissipation in ultraprecision machining, *CIRP Ann.*, 42 (1): 83–86.

Malekian, M.; Park, S.S.; Jun, M.B.G. (2009) Modelling of dynamic micro-milling cutting forces, *Int. J. Mach. Tools Manuf.*, 49 (7–8): 586–598.

Maroju, N.K.; Jin, X. (2019) Mechanism of chip segmentation in orthogonal cutting of Zr-based bulk metallic glass, *J. Manuf. Sci. Eng.*, 141 (8): 081003/1–13.

Maroju, N.K.; Yan, D.P.; Xie, B.; Jin, X. (2018) Investigations on surface microstructure in high-speed milling of Zr-based bulk metallic glass, *J. Manuf. Process.*, 35: 40–50.

Masuzawa, T. (2000) State of the art of micromachining, *CIRP Ann.*, 49 (2): 473–488.

Masuzawa, T.; Tonshoff, H.K. (1997) Three-dimensional micromachining by machine tools, *CIRP Ann.*, 46 (2): 621–628.

Melkote, S.; Kumar, M.; Hashimoto, F.; Lahoti, G. (2009) Laser assisted micro-milling of hard-to-machine materials, *CIRP Ann. Manuf. Technol.*, 58: 45–48.

Mian, A.J.; Driver, N.; Mativenga, P.T. (2011) Chip formation in microscale milling and correlation with acoustic emission signal, *Int. J. Adv. Manuf. Technol.*, 56: 63–78.

Nakayama, K.; Arai, M. (1987) Burr formation in metal cutting, *CIRP Ann.*, 36 (1): 33–36.

Newby, G.; Venkatachalam, S.; Liang, S.Y. (2007) Empirical analysis of cutting force constants in micro-end-milling operations, *J. Mater. Process. Technol.*, 192–193: 41–47.

Ng, C.K.; Melkote, S.N.; Rahman, M.; Kumar, A.S. (2006) Experimental study of micro- and nano-scale cutting of aluminium 7075-T6, *Int. J. Mach. Tools Manuf.*, 46 (9): 929–936.

Niu, Z.; Jiao, F.; Cheng, K. (2018) An innovative investigation on chip formation mechanisms in micro-milling using natural diamond and tungsten carbide tools, *J. Manuf. Process.*, 31: 382–394.

Olvera, O.; Barrow, G. (1996) An experimental study of burr formation in square shoulder face mill-ing, *Int. J. Mach. Tools Manuf.*, 36 (9): 1005–1020.

Palmer, W.B.; Yeo, R.C.K. (1963) Metal flow near the tool point during orthogonal cutting with a blunt tool, *Advances in machine tool design and research: Proceedings of the 4th International M.T.D.R. Conference*, Manchester College of Science & Technology, Manchester, UK, September 1963, 61–71.

Popov, K.B.; Dimov, S.S.; Pham, D.T.; Minev, R.M.; Rosochowski, A.; Olejnik, L. (2006) Micromill-ing: Material microstructure effects, *Proc. I MechE Part B: J. Eng. Manuf.*, 220: 1807–1813.

Ramos, A.C.; Autenrieth, H.; Straub, T.; Deuchert, M.; Hoffmeister, J.; Schulze, V. (2012) Character-ization of the transition from ploughing to cutting in micro-machining and evaluation of the mini-mum thickness of cut, *J. Mater. Process. Technol.*, 212 (3): 594–600.

Ray, D. (2020) *Investigation on Machining Characteristics of Amorphous Bulk Metallic Glass in Mechanical Micro Milling*. PhD diss., National Institute of Technology Durgapur.

Ray, D.; Puri, A.B.; Hanumaiah, N. (2019a) An experimental investigation on the formation of burrs in micro milling of Zr-based amorphous bulk metallic glass, *Int. J. Mater. Prod. Technol.*, 59 (2): 140–171.

Ray, D.; Puri, A.B.; Hanumaiah, N. (2020a) Experimental analysis on the quality aspects of micro-channels in mechanical micro milling of Zr-based bulk metallic glass, *Measurement*, 158 (1): 107622/1–107622/14.

Ray, D.; Puri, A.B.; Hanumaiah, N.; Halder, S. (2019b) Modelling of top burr formation in micro-end milling of Zr-based bulk metallic glass, *J. Micro Nano-Manuf.*, 7 (4): 041004/1–041004/12.

Ray, D.; Puri, A.B.; Hanumaiah, N.; Halder, S. (2020b) Analysis on specific cutting energy in micro milling of bulk metallic glass, *Int. J. Adv. Manuf. Technol.*, 108 (1): 245–261.

Ray, D.; Puri, A.B.; Hanumaiah, N.; Halder, S. (2020c) Mechanistic modelling of dynamic cutting forces in micro end milling of Zr-based bulk metallic glass, *Int. J. Mach. Mach. Mater.*, 22 (6): 527–550.

Schroers, J.; Johnson, W.L. (2004) Ductile bulk metallic glass, *Phys. Rev. Lett.*, 93 (25): 255506/1–255506/4.

Simoneau, A.; Ng, E.; Elbestawi, M.A. (2006) Chip formation during microscale cutting of a medium carbon steel, *Int. J. Mach. Tools Manuf.*, 46 (5): 467–481.

Simoneau, A.; Ng, E.; Elbestawi, M.A. (2007) Modeling the effects of microstructure in metal cutting, *Int. J. Mach. Tools Manuf.*, 47 (2): 368–375.

Son, S.M.; Lim, H.S.; Ahn, J.H. (2005) Effects of the friction co-efficient on the minimum cutting thickness in micro-cutting, *Int. J. Mach. Tools Manuf.*, 45 (4–5): 529–535.

Srinivasa, Y.V.; Shanmugam, M.S. (2013) Mechanistic model for prediction of cutting forces in micro-end milling and experimental comparison, *Int. J. Mach. Tools Manuf.*, 67: 18–27.

Ueda, K.; Manabe, K. (1992) Chip formation mechanism in microcutting of an amorphous alloy, *CIRP Ann.*, 41 (1): 129–132.

Vazquez, E.; Rodriguez, C.A.; Elias-Zuniga, A.; Ciurana, J. (2010) An experimental analysis of pro-cess parameters to manufacture metallic micro-channels by micro-milling, *Int. J. Adv. Manuf. Tech-nol.*, 51: 945–955.

Vogler, M.P.; DeVor, R.E.; Kapoor, S.G. (2004) On the modelling and analysis of machining perfor-mance in micro-end milling, Part I: Surface generation, *J. Manuf. Sci. Eng.*, 126 (4): 685–694.

Waldorf, D.J.; Devor, R.E.; Kapoor, S.G. (1998) A slip-line field for ploughing during orthogonal cut-ting, *J. Manuf. Sci. Eng.*, 120 (4): 693–699.

Waldorf, D.J.; Devor, R.E.; Kapoor, S.G. (1999) An evaluation of ploughing models for orthogonal machining, *J. Manuf. Sci. Eng.*, 121 (4): 550–558. Wang, H.; To, S.; Chan, C.Y.; Cheung, C.F.; Lee, W.B. (2010) A study of regularly spaced shear bands and morphology of serrated chip formation in micro cutting process, *Scr. Mater.*, 63 (2): 227–230.

Wang, W.H.; Dong, C.; Shek, C.H. (2004) Bulk metallic glasses, *Mater. Sci. Eng.*, R 44 (2–3): 45–89.

Weule, H.; Huntrup, V.; Tritschler, H. (2001) Micro-cutting of steel to meet new requirements in min-iaturization, *CIRP Ann.*, 50 (1): 61–64.

Wu, X.; Li, L.; He, N.; Yao, C.; Zhao, M. (2016) Influence of the cutting edge radius and the material grain size on the cutting force in micro-cutting, *Precis. Eng.*, 45: 359–364. Yu, J.; Wang, G.; Rong, Y. (2015) Experimental study on the surface integrity and chip formation in the micro cutting process, *Procedia Manuf.*, 1: 655–662.

Yuan, Z.J.; Zhou, M.; Dong, S. (1996) Effect of diamond tool sharpness on minimum cutting thickness and cutting surface integrity in ultraprecision machining, *J. Mater. Process. Technol.*, 62 (4): 327–330.

Zhang, T.; Liu, Z.; Shi, Z.; Xu, C. (2017) Investigation of size effect on specific cutting energy in mechanical micro-cutting, *Int. J. Adv. Manuf. Technol.*, 91 (5–8): 2621–2633.

Zhong, L.; Liang, L.; Wu, X.; He, N. (2016) Micro cutting of pure tungsten using self-developed polycrystalline diamond slotting tools, *Int. J. Adv. Manuf. Technol.*, 89 (5–8): 2435–2445.

Chapter 2

Effects of Reaction Product Layer Formation in the Fabrication of Cylindrical Tungsten Micro-tools through Micro-ECM Process

Md. Zishanur Rahman, Alok Kumar Das, Somnath Chattopadhyaya, and Md. Quamar Alam

2.1 INTRODUCTION

The demands of parts/components and products with micro-features have increased rapidly in the recent years. Some of the broadly used parts and products containing micro-features include cooling channels, turbine blades, and wire drawing dies. For machining micro-features, methods such as mechanical micromachining, electrochemical micromachining (micro-ECM), electro-discharge micromachining (micro-EDM), electrochemical discharge micromachining (micro-ECDM), photochemical micromachining, and laser-beam micromachining are used (Leo et al., 2014; Anand et al., 2014). Residue stresses develop in micro-parts/products when mechanical micromachining is used due to exerts of significant cutting forces. Thermal residue stresses develop in micro-parts/products when laser micromachining is used. Furthermore, for micromachining, economy of manufacturing cost is essential along with dimensional accuracy and surface quality (Sánchez et al., 2013). Considering these problems, micro-ECM have turned out to be an effective and useful technique for producing micro-features in hard-to-cut materials like tungsten, tool steel, titanium alloys, carbides, tungsten-carbide, and super alloys (Rajurkar et al., 2013). These days, micro-ECM has been getting popularity in the manufacturing of micro-components and micro-parts of numerous products. The concepts of micro-ECM have come up from the conventional electrochemical machining (ECM). In fact, micro-ECM is the application of ECM in micro-fabrication. A basic comparison between micro-ECM and traditional ECM has been presented in Table 2.1 (Bhattacharyya et al., 2004).

The ECM process was made patent in 1929 by Gusseff. After significant advancements around the 1960s, the ECM process became a greater technology in the field of aerospace and aircraft industries for deburring, milling, and finishing operations of big parts/components. These days applications of ECM include various sectors such as aerospace, biomedical, tribology, energy, and automotive (Holstein et al., 2011). The ECM process works on Faraday's laws of electrolysis in which anode (workpiece) and cathode (tool) are immersed in the aqueous electrolyte, kept inside electrolytic cell. Material removal takes place from anodic workpiece at a sufficient potential difference between anode and cathode. The performance ECM process is determined by several process parameters: current density, types of electrolyte, electrolyte concentration, electrolyte flow rate, anodic reactions, and inter-electrode gap (IEG). Machining efficiency, accuracy, and surface finish can be improved by using short pulse current and/or high frequency (Mathew et al., 2012).

DOI: 10.1201/9781003270027-2

Table 2.1 Basic Comparison between Micro-ECM and ECM

Basic Features	Micro-ECM	ECM
Tool-tip size	Micro	Large/medium
Power supply (DC)	Pulsed	Pulsed/continuous
Inter-electrode gap (IEG)	5–50 μm	100–600 μm
Voltage between electrodes	<10 V	10–30 V
Density of current	75–100 A/cm²	20–200 A/cm²
Current flow between electrodes	<1 A	150–10,000 A
Electrolyte flow rate	<3 m/s	10–60 m/s
Concentration of electrolyte	<20 g/l	> 20 g/l
Machining accuracy	± 0.1–0.02 mm	± 0.1 mm
Side gap	<10 μm	> 20 μm
Surface roughness (Ra)	0.4–0.05 μm	1.5–0.1 μm

The control of tool diameter during the fabrication of cylindrical micro-tool is very crucial for micro-ECM process. The dimension accuracy and surface quality of the micro-features obtained by ECM process are directly affected by micro-tool accuracy and its strength, because the job is a mirror image of micro-tool shape (Ghoshal et al., 2013; Reddy et al., 1988; Zhu et al., 1993; Sorkhel et al., 1989). In addition, the micro-tool must have intended characteristics such as extreme stiffness, high thermal and electrical conductivity, corrosion resistance, and good machinability. Tungsten is a preferable micro-tool electrode material due to its properties such as extreme toughness and rigidity, extreme melting point, and chemicals resistance as compared to the other competitive materials (Rahman et al., 2018; Swain et al., 2012).

2.1.1 Passive Layer/Reaction Product Layer Formation and Its Significance

In micro-ECM process, the choice of electrolyte plays a major role as it incidentally determines the electrochemical reactions among the cathodic tool and the anodic workpiece (Rahman et al., 2021a, 2021b). In a few combinations of electrode and electrolyte, the reaction product layer/passive layer (Figure 2.1a and c) is formed on the immersed anode surface. On this basis, the electrolytes are classified as passivating and non-passivating types (Lesch et al., 2011; Klocke et al., 2018; Wang et al., 2017; Liu et al., 2017b; Xu et al., 2016). The non-passivating electrolyte does not form a passive layer on the workpiece during the ECM process. Passive layers are generally composed of metal oxides and metal hydroxide which stands as the barrier for the further electrochemical reaction. Non-passivating electrolytes contain aggressive ions which neutralize the effects of this layer and results in a higher dissolution rate of work material that leads to poor surface finish and machining resolution (Leese et al., 2016). But micromachining with high resolution is very important, especially in micro-tool electrode fabrication. Therefore, the passive electrolyte is preferred to the micro-tool electrode fabrication (Rahman et al., 2021a). Hence, formation of passive layer/reaction product layer during the micro-ECM process with passive electrolyte plays significant role in producing good surface finish with high resolution.

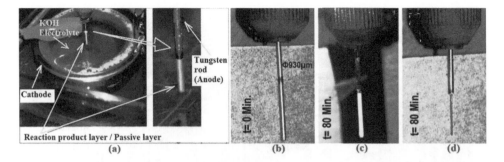

Figure 2.1 (a) Reaction product layer/passive layer. (b) Tool electrode without machining at $t = 0$ minute. (c) Tool electrode at $t = 80$ minutes with deposited reaction product. (d) Micro-tool electrode at $t = 80$ minutes after removal of reaction product.

In this study, the role of reaction product layer formation has been analyzed in the manufacturing of tungsten micro-tool through micro-ECM with KOH electrolyte. Experiments are conducted using L16 mixed-level orthogonal array design.

2.2 EXPERIMENTATION

The in-house built-up micro-ECM setup as depicted in Figure 2.2 is used for fabricating the cylindrical (taper-less) micro-tool. This setup has a rounded-shape machining chamber of a metallic wall (made of stainless steel) seated over a square acrylic plate. A fresh electrolyte is used in each experiment to maintain its uniform pH level. To carry out the tool fabrication process, a tungsten rod of cylindrical shape having 930 μm initial diameters is mounted precisely in the spindle with the help of collet. Prepared KOH(Aq.) electrolytes are poured within the rounded-shape machining chamber (Figure 2.2) and the mounted tungsten rod is positioned static at center position with merged length of 8 mm inside this machining chamber. The power supply (DC) is linked across metallic wall of machining chamber and tool rod to make respectively cathode and anode. The electrolysis starts inside the machining chamber due to potential difference between anode and cathode, and the anodic tool material start to dissolve into the aqueous solution of electrolyte. For controlling the current effects at the lowest edge of anodic tool tip and for fabricating taper-less micro-tool tip, this anodic tool is kept adhered to the acrylic base plate, as shown in Figure 2.2c. Taper-less micro-tool tip have been manufactured using this arrangement which has hardly been appropriately emphasized by other study.

2.2.1 Design of Experiments

For current study, selected machining parameters are DC voltage (volt), concentration of electrolyte (mol/liter), and machining time (minutes). The working range of cutting parameters is determined after many pilot experiments. For conducting the experiments, a mixed-level control factors (Table 2.2) have been decided in which machining time has eight levels, electrolyte concentration and DC voltage have two levels. MRR and tool radius reduction (Δ) are preferred as responses of the experimentation. For accommodating these mixed-level control factors, Taguchi's L16 mixed-level design (as illustrated in Table 2.3) is used to

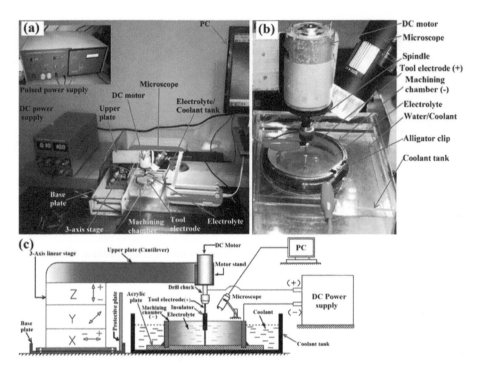

Figure 2.2 (a) Micro-ECM setup. (b) Coolant tank with machining chamber. (c) Setup arrangement for the manufacturing of cylindrical micro-tool.

Table 2.2 Parameters of Experiments and Its Values

Parameters of Experiments	Code	Levels							
		1	*2*	*3*	*4*	*5*	*6*	*7*	*8*
Machining time (minutes)	t	10	20	30	40	50	60	70	80
Concentration of electrolyte (mol/l)	C	3	4						
DC voltage (V)	V	7	11						

conduct the experiments. Each experiment is conducted by emerging 8 mm length of cylindrical tungsten rod inside the KOH (Aq.) electrolyte.

2.2.2 Measurement of Response Output ("Δ" and "MRR")

During this research, response outputs, that is, "MRR" and "Δ" are computed for each micro-tool fabricated using Equations (2.2) and (2.3), respectively. Equations (2.2) and (2.3) are formulated from the geometry of micro-tool (i.e., tool tip) fabricated as depicted in Figure 2.3. To measure the microtool diameters, metallurgical microscope of OLYMPUS (Model: BX51M) is used. A 0.001–10 g high-precision digital weighing scale is used for measuring the weight of workpiece sample after each machining and before each machining.

(i) Formulation of "Δ":

Δ = tool radius before machining–tool radius (average) after machining

Average diameter of micro-tool fabricated: $da = \dfrac{d1 + d2}{2}$ (2.1)

$\Delta = \dfrac{D}{2} - \dfrac{d1 + d2}{4}$ (2.2)

(ii) Formulation of "MRR":

$MRR = \dfrac{w1 - w2}{t}$ (2.3)

Here

D: tool diameter before machining (micrometer)
t: machining time (minute)
L: length of micro-tool tip fabricated = length between $d1$ and $d2$ (i.e., 7,000 μm)
$d1$: tool diameter after machining at free end (micrometer)
$d2$: tool diameter after machining at L distance (micrometer), as shown in Figure 2.3c
$W1$: weight of workpiece before machining (microgram)
$W2$: weight of workpiece after machining (microgram)

2.3 RESULTS ANALYSIS

This section presents experimental results and their analysis using Taguchi design. Table 2.3 illustrates the results of experiments. Figure 2.4 illustrates the fabricated micro-tools corresponding to each experimental run. To analyze the effects of cutting parameters on "MRR" and "Δ", main effects plot of mean are generated (Figures 2.6 and 2.7).

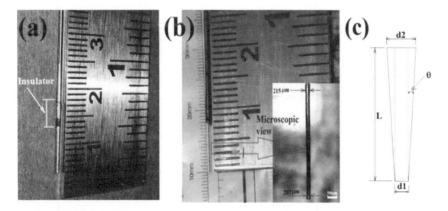

Figure 2.3 (a) Cylindrical rod of tungsten material before machining. (b) Fabricated micro-tool tip at experiment number 8. (c) Fabricated tool-tip geometry.

Table 2.3 Results of Experiments Corresponding to Each Experimental Run

Exp. Run	Cutting Parameters and Levels			Designation	W1 (mg)	W2 (mg)	W1–W2 (mg)	t (min)	MRR (µg/min)	$\frac{D}{2}$ (µm)	d1 (µm)	d2 (µm)	$\frac{d1+d2}{4}$ (µm)	Δ (µm)
	t	C	V											
1	1	1	1	t1C1V1	320	295	25	10	2,600	465	866	867	433.25	31.75
2	1	2	2	t1C2V2	310	278	32	10	3,200	465	740	742	370.50	94.50
3	2	1	1	t2C1V1	310	268	42	20	2,100	465	733	738	367.75	97.25
4	2	2	2	t2C2V2	282	230	52	20	2,600	465	607	611	304.50	160.50
5	3	1	1	t3C1V1	304	252	52	30	1,733	465	618	624	310.50	154.50
6	3	2	2	t3C2V2	290	224	66	30	2,166	465	497	501	249.50	215.50
7	4	1	1	t4C1V1	320	259	61	40	1,525	465	520	526	261.50	203.50
8	4	2	2	t4C2V2	290	215	75	40	1,875	465	415	419	208.50	256.50
9	5	1	2	t5C1V2	277	207	70	50	1,400	465	439	448	221.75	243.25
10	5	2	1	t5C2V1	279	199	80	50	1,600	465	350	359	177.25	287.75
11	6	1	2	t6C1V2	276	198	78	60	1,300	465	375	386	190.25	274.75
12	6	2	1	t6C2V1	270	184	86	60	1,433	465	294	305	149.75	315.25
13	7	1	2	t7C1V2	260	174	86	70	1,228	465	319	330	162.25	302.75
14	7	2	1	t7C2V1	376	282	94	70	1,342	465	247	259	126.50	338.50
15	8	1	2	t8C1V2	257	163	94	80	1,175	465	274	286	140.00	325.00
16	8	2	1	t8C2V1	340	239	101	80	1,262	465	207	220	106.75	358.25

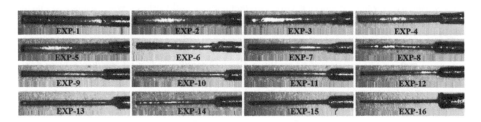

Figure 2.4 Fabricated micro-tools corresponding to each experimental run.

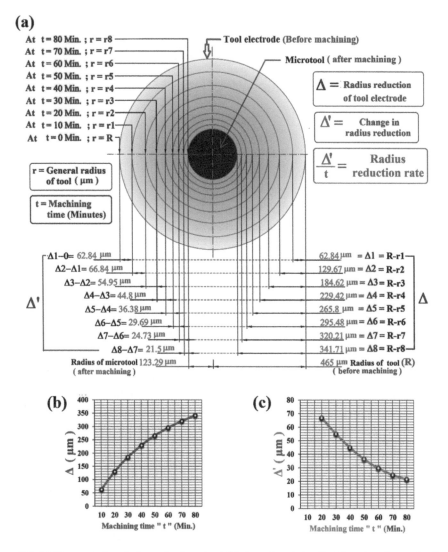

Figure 2.5 (a) Cross-sectional view of micro-tool fabricated and details of tool radius-reduction Δ with respect to machining time t. (b) Graph between Δ and t. (c) Graph between Δ′ and t.

Table 2.4 Recorded Value of Current Flow between Anode and Cathode with respect to Machining Time *t* (Minute)

Exp. No. ➡	7	8	15	16
Parameters ➡	C: 3 mol/l; V: 7 V	C: 4 mol/l; V: 11 V	C: 3 mol/l; V: 11 V	C: 4 mol/l; V: 7 V
T (minute) ⬇	I (A)	I (A)	I (A)	I (A)
Start "s"	0.52	0.70	0.55	0.64
5	0.10	0.13	0.08	0.12
10	0.07	0.09	0.07	0.08
15	0.06	0.06	0.06	0.06
20	0.05	0.05	0.05	0.05
25	0.04	0.04	0.04	0.04
30	0.03	0.03	0.03	0.03
35	0.02	0.02	0.02	0.02
40	0.02	0.02	0.02	0.02
45			0.01	0.01
50			0.01	0.01
55			<0.01	<0.01
60			<0.01	<0.01
65			<0.01	<0.01
70			<0.01	<0.01
75			<0.01	<0.01
80			<0.01	<0.01

2.3.1 Characteristics of Reaction Product (Passive Layer) Formed and Its Effects on Tool Radius Reduction and MRR

During the tungsten cylindrical micro-tool electrodes fabrication using micro-ECM process using KOH electrolyte, it has been observed that the electrochemical dissolution in anodic tool electrode takes place with formation of reaction product (passive layer) on it (Figures 2.1 and 2.2). According to Leese et al. (2016), passive layers are generally composed of metal oxides and metal hydroxide which stands as the barrier for the further electrochemical reaction. It means formation of passive layers either stops the further electrochemical reaction or reduces the electrochemical reaction rate and depends on characteristics of formed reaction product layer/passive layers. This section deals with characterization of reaction product (passive layer) formed on the anodic tungsten tool electrode and its effects on tool radius reduction rate during the tungsten cylindrical micro-tool electrodes fabrication using micro-ECM process using aqueous KOH electrolyte.

During all experiments, it is found that the reaction product (passive layer) formed on the anodic tungsten tool is insoluble in KOH(Aq.) electrolyte and this layer is deposited on the micro-tool electrode throughout the machining time, as depicted in Figure 2.1. Figure 2.5 depicts the cross-sectional view of the micro-tool fabricated and details of tool radius reduction during the process. This plot is drawn by using the results obtained for FIT values of "Δ". In this figure t is the machining time (minute), r is the general radius of tool (μm), R is the maximum radius of tool electrode before machining (μm), Δ is the reduction in tool radius (μm), and $Δ'$ represents the change in tool radius reduction (μm) after each 10 minutes. Figure 2.5b shows that reduction in tool radius (Δ) increases with rise of t, which indicates that thickness of passive/reaction product layer increases with the increase of t. Current flow rate between the anode and cathode has been recorded, as shown in Table 2.4, which indicates that current flow rate decreases as the thickness of passive/reaction product layer increases with the increase of t. It indicates that the formation of reaction product (passive layer) on the tool electrode treats as barrier, therefore material removal rate (MRR) decreases (Figure 2.6). Hence the tool radius reduction rate decreases (Figure 2.5c) and, therefore, fabrication in tungsten micro-tool with aqueous KOH electrolyte takes long machining time. However, during the tungsten cylindrical micro-tool fabrication through micro-ECM process with aqueous KOH, it has been found that the formation of reaction product/passive layer performs an important role in control of uniform dissolution of tungsten tool electrode (anode) throughout its immersion depth and generates a uniform cylindrical micro-tool (Figure 2.1d). Since the reaction product layer of tungsten tool electrode is not soluble in KOH electrolyte (Figure 2.1c), hence formation of reaction product layer can help in controlling the machining rate.

2.3.2 Effects of Cutting Parameters on MRR and Tool Radius Reduction (Δ)

The main effects plot of MRR (Figure 2.6) depicts that the machining time (t) plays more significant role on MRR, and MRR decreases with increment of t because of an increase in the radial thickness of the reaction product/passive layer. This figure also depicts that MRR increases with the increment of concentration of electrolyte C and DC voltage V both. Figure 2.7 illustrates the main effects plot of "Δ". This figure depicts that the machining time (t) plays more significant role on Δ, and "Δ" increases with rise of machining time as well as concentration of electrolyte and DC voltage.

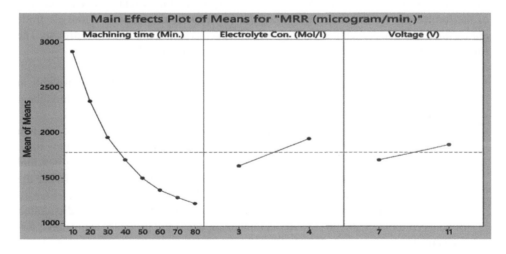

Figure 2.6 Main effects plot of mean for MRR.

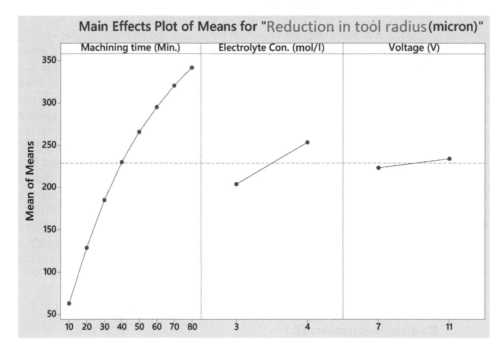

Figure 2.7 Main effects plot of mean for tool radius reduction.

2.4 CONCLUSIONS

This chapter emphasizes the effects of reaction product layer formation in the manufacturing of tungsten cylindrical micro-tools through micro-ECM with KOH(Aq.). The effects of reaction product layer formation have been analyzed on the two responses: material removal rate (MRR) and tool radius reduction (Δ). This study also predicts the current flow characteristics of the formed passive/reaction product layer on the anodic tungsten micro-tool electrode during the microelectrochemical etching with an aqueous solution of KOH electrolyte. The following are the conclusions based on results of experiments and their analysis:

- Reaction product layer/passive layer formed on the tungsten micro-tool is insoluble in KOH liquid electrolyte and this layer is deposited on micro-tool electrode throughout the machining time.
- Current flow rate between anode and cathode decreases as the thickness of reaction product/passive layer on the anodic tool increases with the increase of machining time. It indicates that the formation of reaction product/passive layer on the tungsten tool electrode treats as barrier, therefore material removal rate (MRR) decreases. Hence the tool radius reduction rate decreases; as a result, it takes long machining time.
- Formation of reaction product layer plays crucial role in control of uniform dissolution of tungsten tool (anode) throughout its immersion depth and generates appreciable surface quality.

- With the increase of machining time, reduction in tool radius increases, while MRR decreases because of the increase in thickness of reaction product layer on tool.

REFERENCES

Anand, R.S.; Patra, K. Modeling and simulation of mechanical micro-machining—A review. *Machining Science and Technology* 18, no. 3 (2014): 323–347.

Bhattacharyya, B.; Munda, J.; Malapati, M. Advancement in electrochemical micro-machining. *International Journal of Machine Tools & Manufacture* 44 (2004): 1577–1589.

Ghoshal, B.; Bhattacharyya, B. Influence of vibration on micro-tool fabrication by electrochemical machining. *International Journal of Machine Tools and Manufacture* 64 (2013): 49–59.

Holstein, N.; Krauss, W.; Konys, J. Development of novel tungsten processing technologies for electrochemical machining (ECM) of plasma facing components. *Fusion Engineering and Design* 86, nos. 9–11 (2011): 1611–1615.

Klocke, F.; Harst, S.; Ehle, L.; Zeis, M.; Klink, A. Surface integrity in electrochemical machining processes: An analysis on material modifications occurring during electrochemical machining. *Proceedings of the Institution of Mechanical Engineers, Part B: Journal of Engineering Manufacture* 232, no. 4 (2018): 578–585.

Leese, R.J.; Ivanov, A. Electrochemical micromachining: An introduction. *Advances in Mechanical Engineering* 8, no. 1 (2016): 1687814015626860.

Leo Kumar, S.P.; Jerald, J.; Kumanan, S.; Prabakaran, R. A review on current research aspects in tool-based micromachining processes. *Materials and Manufacturing Processes* 29, nos. 11–12 (2014): 1291–1337.

Lesch, A.; Wittstock, G.; Burger, C.; Walther, B.; Hackenberg, J. External control of anodic dissolution mechanisms of 100Cr6 in nitrate/chloride mixed electrolytes. *Journal of Electrochemical Science and Engineering* 1, no. 1 (2011): 39–54.

Liu, W.; Ao, S.; Li, Y.; Liu, Z.; Wang, Z.; Luo, Z.; Wang, Z.; Song, R. Jet electrochemical machining of TB6 titanium alloy. *The International Journal of Advanced Manufacturing Technology* 90, no. 5 (2017a): 2397–2409.

Liu, W.; Ao, S.; Li, Y.; Liu, Z.; Zhang, H.; Manladan, S.M.; Luo, Z.; Wang, Z. Effect of anodic behavior on electrochemical machining of TB6 titanium alloy. *Electrochimica Acta* 233 (2017b): 190–200.

Mathew, R.; Sundaram, M.M. Modeling and fabrication of micro tools by pulsed electrochemical machining. *Journal of Materials Processing Technology* 212, no. 7 (2012): 1567–1572.

Rahman, Z.; Kumar Das, A.; Chattopadhyaya, S. Microhole drilling through electrochemical processes: A review. *Materials and Manufacturing Processes* 33, no. 13 (2018): 1379–1405.

Rahman, Md Z.; Kumar Das, A.; Chattopadhyaya, S. Effects of balance electrode in deep micro-holes drilling in "nickel plate" through μECM process using H_2SO_4 electrolyte. *Materials Today: Proceedings* 43 (2021a): 1431–1436.

Rahman, Md Z.; Kumar Das, A.; Chattopadhyaya, S. Fabrication of high aspect-ratio tungsten microtools through controlled electrochemical etching. *Materials and Manufacturing Processes* 36, no. 11 (2021b): 1236–1247.

Rajurkar, K.P.; Sundaram, M.M.; Malshe, A.P. Review of electrochemical and electrodischarge machining. *Procedia CIRP* 6 (2013): 13–26.

Reddy, M.S.; Jain, V.K.; Lal, G.K. Tool design for ECM: Correction factor method. *Transactions of the ASME* (1988): 111–118.

Sánchez, J.A.; Plaza, S.; Gil, R.; Ramos, J.M.; Izquierdo, B.; Ortega, N.; Pombo, I. Electrode set-up for EDM-drilling of large aspect-ratio micro holes. *Procedia CIRP* 6 (2013): 275–280.

Sorkhel, S.K.; Bhattacharyya, B. Computer aided design of tools in ECM for accurate job machining. *Proceedings of the ISEM-9, Japan* (1989): 240–243.

Swain, A.K.; Sundaram, M.M.; Rajurkar, K.P. Use of coated microtools in advanced manufacturing: An exploratory study in electrochemical machining (ECM) context. *Journal of Manufacturing Processes* 14, no. 2 (2012): 150–159.

Wang, D.; Zhu, Z.; He, B.; Ge, Y.; Zhu, D. Effect of the breakdown time of a passive film on the electrochemical machining of rotating cylindrical electrode in $NaNO_3$ solution. *Journal of Materials Processing Technology* 239 (2017): 251–257.

Xu, Z.; Chen, X.; Zhou, Z.; Qin, P.; Zhu, D. Electrochemical machining of high-temperature titanium alloy Ti60. *Procedia CIRP* 42 (2016): 125–130.

Zhu, D.; Yu, C.Y. Investigation on the design of tool shape in ECM. *ASME-Publications-PED* 58 (1993): 181–190.

Chapter 3

Hybrid Additive Manufacturing

Arun Sharma, Aarti Rana, and Dilshad Ahmad Khan

3.1 WHY DO WE NEED HYBRID ADDITIVE MANUFACTURING?

In today's ultra-competitive market, the only way to stay afloat is to provide at a minimal price, the highest-quality standard product with all conceivable features. Although additive manufacturing (AM) brings ample advantages, there are a large number of issues encountered in this advancement. It would necessitate additional innovative activities well before this advancement could be implemented in different industries. Let us put light on a few issues: in additive manufacturing, voids are created in between different consecutive layers of material produced because of a decrease in interfacial bonding (adhesion) between built components/layers throughout the production process. This results in greater porosity which leads to further compromises on mechanical performance. Another eye-catching hurdle for AM is its anisotropic behavior which results in different mechanical characteristics, especially compressive or tensile nature in vertical direction, as opposed to the behavior of AM parts in the horizontal direction.

Additionally, due to the concept of tessellation in computer-aided design, converting a model into a three-dimensional build part leads to ambiguity and errors. This is more prevalent in curved surfaces. Further, AM printing in applications such as aviation, toys for toddlers, and architecture is undesirable (Ngo et al., 2018), Other major issues with additive manufacturing are high production costs, the limited size of the component, and post-processing. In this chapter, different HAM processes will be discussed so that the aforementioned AM problems can be fixed or mitigated to some extent.

Products produced by AM need both pre-processing and post-processing which hampers the cost, speed, and efficiency. Further, surface roughness and surface texture decide how long the part will work in certain cases, but production by AM leads to poor surface quality. This again brings the need for HAM.

3.2 HYBRID MANUFACTURING

Prior to understanding what hybrid additive manufacturing (HAM) is, it is very much important to understand hybrid manufacturing. Hybrid manufacturing has been around for a long time as a way to improve part quality and production by combining two or more processes. Several hybrid approaches have been described using this phrase in the literature such as hybrid process, hybrid machine, hybrid material, hybrid structure, or hybrid functionality. The same is shown in Figure 3.1 and explained in the forthcoming sections.

DOI: 10.1201/9781003270027-3

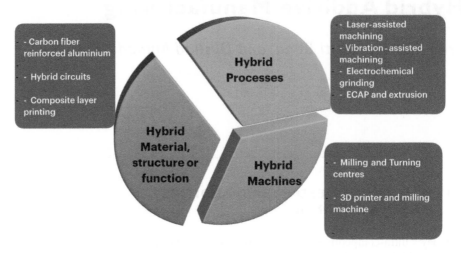

Figure 3.1 Different hybrid approaches.

3.2.1 Hybrid Process

A hybrid process is characterized as simultaneous generating phenomena that would normally be induced by different processes (Nau et al., 2011). CIRP (2011) advocated unrestricted and restricted definitions of hybrid manufacturing processes in 2010. These definitions are as follows:

> ***Unrestricted definition:*** As per this definition, a hybrid process combines more than one existing production process into a replacement integrated setup, permitting the advantages of every separate process to be leveraged interactively or collaboratively, so that each process amplifies the impact of the other. This is based on the "$1 + 1 = 3$" effects.
>
> ***Restricted definition:*** According to this definition, hybrid process involves the simultaneous application of multiple treatment principles whether physical, chemical, or controlled to the same treatment area.

Figure 3.2 depicts how a hybrid process would possibly enhance manufacturing with the aid of reduced process chains or realizing new product attributes. Thanks to the stated technology integrations, now it is possible to shorten the process chain. A single hybrid process can replace several conventional processes (CP), as can be seen in Figure 3.2 (CP-2 and CP-3 combined to get a hybrid process). Regardless of the fact that a hybrid process may be more complex and need more process stability, this might contribute to a decrease in the number of planning activities necessary to synchronize several processes in a process chain. However, achieving new product attributes is a secondary goal of hybrid process development.

Hybridization, for example, can enable a traditional process to mill novel materials that could not previously be machined by the traditional process due to technological limitations.

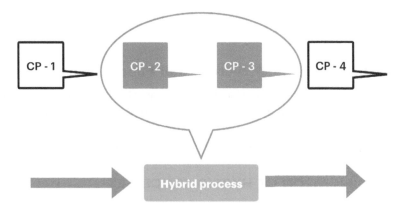

Figure 3.2 Reduction in process chain by hybridization of the processes.

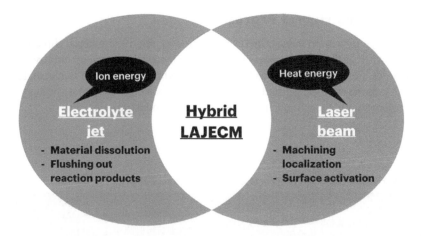

Figure 3.3 Development of hybrid process by combining different processes to generate a third better process.

Innovative gadgets that could not be created because of the high-tech limits of contemporary manufacturing technology are probably realized with the assistance of these new production capabilities. The "1 + 1 = 3" effect occurs when the sum of the hybrid processes exceeds the total of the separate processes. (Sealy et al., 2018).

Figure 3.3 shows how the significant properties of an electrolyte jet (say material dissolution, flushing out reaction products, etc.) and a laser beam (machining localization and surface activation) combined together to develop hybrid LAJECM.

3.2.2   Hybrid Machine

A hybrid process is not the same as a hybrid machine. A hybrid machine is a device that integrates various production processes into one such as a machining center that combines milling and turning or a combination of a milling machine along with a 3D printer. The most important thing to know about hybrid machines is that they refer to the foundation of the machine rather than the processes that make it up.

3.2.3   Hybrid Material, Structure, and Functionality

Hybrid materials, structures, and functionalities are created by the combination of more than one material to generate a hybrid component, structure, or functionality. The characteristics of a hybrid material should be modified or replaced totally. Examples include hybrid circuits, carbon fiber reinforced aluminum, and composite layer printing.

3.3   CLASSIFICATION OF HYBRID PROCESSES

The hybrid processes can be classified into two categories: assisted hybrid processes and mixed hybrid processes (Lauwers et al., 2014).

3.3.1   Assisted Hybrid Process

In assisted-type hybrid processes, the primary process removes material directly from the workpiece, whereas the lateral process just helps the material removal by improving machining conditions. For example, various laser-assisted hybrid processes in Lee et al. (2016).

Table 3.1 shows various assisted hybrid processes as a combination of primary and secondary processes and their frequency of occurrence in industrial use.

Table 3.1 Various Assisted Hybrid Processes (Lauwers et al., 2014)

Combination of assisted hybrid processes in ultrasonic process — Very frequent - ▲, Frequent - ●, Partly - ◆		Primary processes										
		Turning	Milling	Drilling	Grinding	Polishing	EDM	ECM	Laser	Forming	Shearing	Etching
Secondary processes	Vibration assisted	▲	◆	●	▲	●	▲	●	◆	◆		
	Laser assisted	▲	●	●	◆		◆			●	◆	◆
	Water jet assisted								◆			
	Pressure fluid assisted	●			◆					▲		
	Magnetic assisted	◆				●	◆	◆		◆		
	Conductive heat assisted								●			

The following are some of the examples of assisted hybrid processes:

(a) *Vibration-assisted hybrid machining processes:* In these processes, vibration aids removal of the material, removal of by-product/waste, and disposal of waste (Brehl et al., 2008).
(b) *Laser-assisted machining:* In laser-assisted machining, the laser assists the machining process by softening the workpiece (Brecher et al., 2011).
(c) *Media-assisted manufacturing:* In these processes, the primary process is assisted by coolant or lubricant (Wang et al., 2000).

3.3.2 Combined/Mixed-Type Hybrid Processes

In these processes, all the corresponding processes, AM process, as well as the lateral process, take a major part in the material removal. Some examples are (a) combining EDM or ECM with grinding (Koshy et al., 1997) and (b) extrusion in combination with equal channel angular pressing (Matsubara et al., 2003).

3.4 HYBRID ADDITIVE MANUFACTURING

HAM can be classified into the following three subunits: hybrid additive manufacturing processes, hybrid additive machines, and hybrid additive materials. The capabilities of HAM add a new perspective to design by defining new material design rules for the AM framework (Sealy et al., 2018).

3.4.1 HAM Process

The HAM process can be described as the combination of AM process with at least one fully integrated and synergistically combined lateral process or source of energy. This combination enhances the quality of the component, its function, and/or performance of the process. Generally, subtractive processes like machining or surface improving processes like laser shock peening (LSP) or burnishing (Gibson et al., 2021) are used as lateral processes. The significant terms related to the HAM have been discussed in the following sections.

3.4.1.1 Fully Integrated Processes

These processes are completely interconnected. By integration, it means the primary and the secondary processes cannot be separated from each other during the build of the components. Pre- and post-printing processes are not considered linked or integrated. This full coupling of the two processes helps in distinguishing pre-processing and post-processing from the hybrid AM process. Pre-processing steps take place before layer assembly. After the build is finished, post-processing is done. Post-processing in additive manufacturing is essential to enhance properties like surface finish. Sometimes post-processing is used to give an aesthetic touch to the built product.

3.4.1.2 Synergy

By synergy, it means the primary process, that is, AM process and the secondary process such as machining, laser-assisted, and rolling work together, either concurrently (as in laser-assisted plasma deposition [LAPD]) or as part of a cyclical process chain. Almost all of the HAM processes are cyclic in nature and hence are also called sequential processes. Secondary processes can be used after each layer or after a particular set of layers as per the requirements. Machining the build component after each layer or after a particular set of layers to make them smooth for the next layer can be an example of a sequential process.

3.4.1.3 Improvements to Parts and/or Processes

The third subcategory of hybridization is associated with the improvement of the part and/or process. This refers to how the HAM process affects the parameters such as the quality and functionality of the component, or the performance of the process. It should be noted that the bulk of lateral operations in HAM does not support the printing process, unlike hybrid manufacturing techniques. Rather, the portion of its functionality is the key benefit. Different hybrid additive processes are discussed later on in this chapter.

3.4.2 Hybrid AM Machines

Applying additive and subtractive processes in series is one of the most prevalent hybrid manufacturing approaches. The machine might, for example, 3D print a near-net form component that would otherwise be cast or forged. Then it might be used to finish this part with its subtractive capabilities. The most typical integrated platforms make use of milling and directed energy deposition technology. Sheet lamination, powder bed fusion (PBF), and material extrusion are the other additive manufacturing technologies (AMTs) that can be made use of in HAM machines. Table 3.2 shows different hybrid additive manufacturing machines along with primary and secondary processes.

3.4.3 HAM of Materials, Structures, or Functions

Another outstanding capability of HAM is to print a part with more than one material taken together. Moreover, it can build a single component with multiple structures as

Table 3.2 Different HAM Machines along with the Primary and Secondary Processes [14]

Additive Process	Product, Company	Subtractive Process
Sheet lamination	Formation, Fabrisonic	3-axis CNC machining
Direct energy deposition	LASERTEC 65 3D, DMG Mori Selki	5-axis CNC machining
	NT 4300 3D, DMG Mori Selki	5-axis CNC machining, Turning
	WFL Millturn Technologies	5-axis CNC machining, Turning
Cold spraying	MPA 40, Hermle AG	5-axis CNC machining
Powder bed fusion	Lumex Avance—25, Matsuura Machinery Corp. OPM250E, Sodick	3-axis CNC machining
Material jetting	Solidscape Product Lines, Solidscape Inc. (Stratasys)	Planar milling

well. When fabricating a single part, multi-material production employs many materials. Color, density, texture, microstructure, and other properties of the materials also can be varied. The aim of combining or integrating dissimilar materials together is to enhance the functional capability of the product to meet the growing demand for high performance and high quality in this industrial era. The functional complexity of additively manufactured parts can be hybridized. That is, multiple-purpose functional devices are built into a single unit.

UC Berkeley, for example, printed integrated electronic parts such as wireless sensors, inductors, capacitors, and resistors. The technique is used to create the "smart cap" which aids in identifying deterioration in the milk cartoon, etc. (Wu et al., 2015; Wu et al., 2020).

3.5 CLASSIFICATION OF HAM PROCESSES BASED ON SECONDARY PROCESSES

HAM processes can be classified on the basis of the secondary process used to assist or work simultaneously with the primary process. This classification, however, failed and is not in accordance with the prescribed criteria of hybrid machining (i.e., sequential processes do not come under hybrid processes). The classification based on secondary processes has been disused as follows:

(a) Machining-based HAM: Machining is the most prevalent process to be combined with AM to make the process hybrid with the primary purpose of improving surface finish and geometrical precision in AM processes (Flynn et al., 2016), etc.

(b) Thermal-based HAM: The second most prevalent type of hybrid AM process is in which thermal processes are combined with AM processes. In these processes, thermal energy is used to recondition the quality of the material of the already deposited layers. It can also be used to improve printing. Surface treatments like erosion or laser remelting (Yasa et al., 2011a, 2011b) and HAM by laser-assisted melting (Qian et al., 2008) fall in this category.

(c) Mechanical surface treatments–based HAM: In these processes, the mechanical surface treatment processes are combined with the AM processes. The mechanical surface treatment methods reform the already build layer. This is done to achieve an enhanced surface finish, finer microstructure, and higher hardness. These can also be used to reduce distortion, stress relief, and improvement in the density of the component. Further, these processes are also used to obtain favorable compressive residual stresses. HAM by peening (Sealy et al., 2016) or HAM by rolling (Zhou et al., 2016) fall into this category, which are discussed in detail in the forthcoming section.

(d) Solid-state stirring–based HAM: Solid-state stirring–based HAM processes are a less-explored type of HAM process. By combining friction stir processing (FSP) with AM process, the microstructure of the build can be refined. This in turn results in improved mechanical qualities.

Some of the HAM processes falling in the aforementioned categories have been discussed in detail in the following sections and depicted in Figure 3.4.

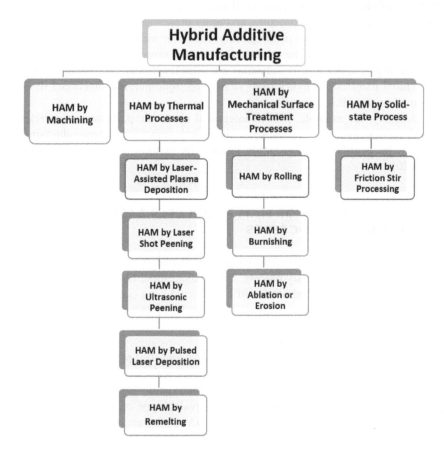

Figure 3.4 Different hybrid AM processes.

3.5.1 HAM by Machining

In HAM literature, machining is known as the most common secondary process (Hur et al., 2002). In this, machining techniques like milling or turning are coupled with AM process to bring the desired product. Table 3.3 shows different HAM processes hybridized by combining AM and machining processes.

Near-net-shape items are produced using AM, whereas machining at successive layer intervals improves surface finish and geometrical precision. Some of the additive technologies that have been employed in combination with machining include sheet lamination, plasma deposition, laser deposition, laser cladding, and laser melting (Sealy et al., 2018).

It is worth noting that post-processing after printing does not call by HAM processing because it does not fit the fully coupled condition. However, for complex geometries or interior features, machining is frequently necessary between layer intervals and would be called a HAM process because separate machining and printing in such a process is almost not possible as it leads to an increase in production cost and lead times. The main machining

Table 3.3 Different HAM Processes Hybridized by Combining AM and Machining Processes [25–34]

	AM Process Category	Material Feedstock	Type of Material	Material Distribution
1. Laser				
(A) Laser welding + machine	DED	Powder	Metal	Deposition nozzle
(B) Laser cladding + milling	DED	Powder	Metal	Deposition nozzle
(C) Laser deposition + milling	DED	Powder	Metal	Deposition nozzle
(D) Selective laser sintering + milling	PBF	Powder	Metal	Powder bed
2. Plasma/arc				
(A) 3D welding + milling	DED	Wire	Metal	Deposition nozzle
(B) Plasma deposition + milling	DED	Powder	Metal	Deposition nozzle
(C) Micro-casting + milling + shot peening	DED	Powder	Metal	Deposition nozzle
3. Solid-state fusion				
(A) Ultrasonic welding + milling	Sheet lamination	Sheet	Metal	Sheet stack
(B) Layered compaction manufacturing + milling + sintering	PBF	Powder	Ceramic	High-density green compact

Source: (Choi et al., 2001; Jeng et al., 2001; Fessler et al., 1996; Ichimura et al., 2014; Akula et al., 2006; Xinhong et al., 2010; Merz et al., 1994; Friel et al., 2013)

technique used in this aspect is milling. Milling in AM is used for two purposes: one is to strengthen the printed component's surface finish, and the second is to keep the consistent thickness of the layer and to bring a smooth surface for subsequent printing. An end mill cutter is used to improve the surface quality of the sidewall solely on a single or even more layers. Face milling is used to generate a smooth, fresh surface for continued printing (Akula et al., 2006; Xinhong et al., 2010). Karunakaran et al. reported that in metal arc additive manufacturing, the elimination of an oxidized layer during face milling impacted the build process (Karunakaran et al., 2008). Because the oxide layer was eliminated, the arc (Table 3.3) and the weld bead were more constant. This example is advantageous in the sense that a non-simultaneous lateral process can have an indirect effect on the primary print process.

Direct energy deposition (DED) is a frequently hybridized AM process with machining. In fact, DED technology is used in most HAM machines available. In DED, a heat source melts a powdered or wire-like material on a substrate in the form of subsequent layers. These processes take the benefit of increased versatility provided by 5-axis or 7-axis milling machines to machine the subsequently deposited layers. On non-planar surfaces, to machine and deposit material, DED systems are preferred. This property of DED systems makes them a preferable choice to repair turbine blades. In DED, the high-velocity powder is dispersed. The other AM process that can be used is PBF. Using a laser or an electron beam, onto a substrate, thin layers of powder are begun to melt in PBF. Until sintered or melted, powder in this system remains stagnant in a bed. In sheet lamination, at ultrasonic frequencies, using oscillatory shear forces, sheets of metal are piled and fused together. In the solid-state fusion process, between both surfaces, by developing a powerful metallurgical bond, coalescence is achieved. To achieve precise dimensions, milling of the deposited layer is done after one or more layers have been built.

Hybrid AM by machining poses distinct hurdles in terms of dimensional variation and machining accuracy. Tool path planning can be difficult due to distortion produced by localized heating (Wang et al., 2015). Furthermore, frequent changeover between machining and printing is costly. PBF and machining are being combined in several industrial applications. For small components, 3-axes milling machine by Coherent Laser, available by the name Creator Hybrid PBF, can be used. Coherent is working on solutions for both laser powder bed fusion (L-PBF) and direct metal deposition (DMD). Matsuura, a company dealing in HAM machines, has released the Avance-25 and Avance-60 hybrid PBF and milling machines. Both of these are medium-sized (Gibson et al., 2021).

3.5.2   HAM by Rolling and Burnishing

Rolling HAM (RHAM) fixes two major problems that were encountered during AM. In metal additive manufacturing, for example, overlapping grains or layers cause dimensional anomalies in the printed component. While machining may correct these flaws, wastage of material in the form of cutting chips raises production costs. Instead of eliminating material, some of these errors can be minimized by rolling. Second, during AM, the end product is deformed or stretched due to the generation of residual stress. RHAM, without any material loss, accomplishes geometrical accuracy and stress relief. These internal stresses can be mitigated by surface treatment processes like peening.

Colegrove et al. (Colegrove et al., 2017) made use of metal-arc AM (particularly WAAM) and employed slotted and curved rolling tools. Integrating rolling after every layer leads to a reduction in distortion, grain refinement improvement, and improvement in mechanical characteristics. Hybrid AM rolled samples had greater maximum strength, hardness, and elongation than as-cast material. Furthermore, rolling the samples lowered tensile stresses.

Zhou et al. employed a piece of metamorphic hot-rolling equipment that could roll a component on a single, two, and even three sides. The equipment has one roller in the horizontal direction and two vertical rollers. The former roller performs upon the uppermost flat surface of the workpiece, while the latter rollers operate on vertical surfaces. The rollers aid in the improvement of thin-wall systems. Hot rolling resulted in an enhanced tensile strength of about 33% over a standard sample, refined grain structure, and better geometrical correctness as per the research findings (Zhou et al., 2016).

Burnishing is the technique of compacting the surface of the metal by rubbing it with a tiny hard tool. It is a surface treatment procedure similar to rolling that improves surface roughness and residual stress. It also helps in improving microstructure and hardness. A rolling/sliding tool which is either a ball or a roller is used for burnishing. The tool slides/rolls across the material's surface. This causes persistent distortion in a tiny section of the surface. This results in the flow of the material. Moreover, on the surface, the peaks and valleys become almost negligible. Functionally graded attributes with better geometrical components are achievable when burnishing occurs between printed layers.

The benefits of rolling/burnishing for different mechanical characteristics such as corrosion, fatigue, and wear cannot be completely realized without a sequential process chain that includes rolling or burnishing. As a result, in hybrid additive manufacturing, these processes must be fully integrated in order to have a synergistic effect on the performance of the component.

3.5.3   Friction Stir Additive Manufacturing

Friction stir additive manufacturing (FSAM) employs friction stir welding (FSW) to irreversibly fuse two surfaces. A non-consumable spinning tool is introduced into a workpiece in FSAM. The tool contains a pin and a shoulder. It is made up of refractory material (Palanivel et al., 2015). The tool is passed over the part's surface. Due to this, heat is generated and a significant amount of permanent deformation takes place. By merging the highly plasticized material, two layers are further connected. Unlike FSW, there is no melting of the workpiece to induce coalescence. This usually results in recrystallization and significant grain refinement in metals. The fatigue and mechanical performance of these metallurgical advancements have accelerated. The schematic diagram of FSAM is shown in Figure 3.5. FSW has also been tested in polymers and composites with great success.

A similar strategy can be employed in the case of FSAM as well for polymers and composites. FSAM is a sequential HAM technique. To enhance the properties of objects made using other AM techniques like direct energy deposition or material extrusion such as metallurgical, mechanical, and chemical, friction stir processing can be used. Francis et al. (Francis et al., 2016) looked at the effects of FSP on Ti-6Al-4V parts made via DED. The grains were cleansed and the grain hardness was raised. Based on fatigue evaluations of comparable microstructures, the authors believe that fatigue life might be enhanced.

3.5.4 HAM by Remelting

In this process, previously fused material is remelted using an energy source. The source can be an electron beam or a laser, as shown in Figure 3.6. Even with the use of a laser, the laser strengths are normally low, preventing evaporation. In other words, no material is removed but rather altered.

To better fuse layers together, remelting elevates previously placed material to the melting point. The melted material fills the pores generated during 3D printing, resulting in a component density of above 99%. Remelting can come into play following each layer or after a series of layers. The changes in the mechanical, physical, and chemical characteristics of the remelted region of a particular material take place. These changes are dependent on various parameters like laser scanning speed, laser power, and scanning pattern. The

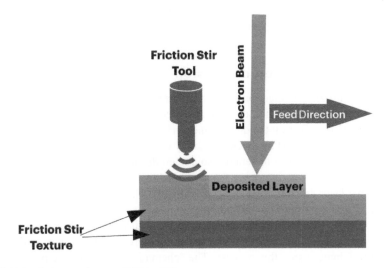

Figure 3.5 Hybrid electron beam—PBF by FSP.

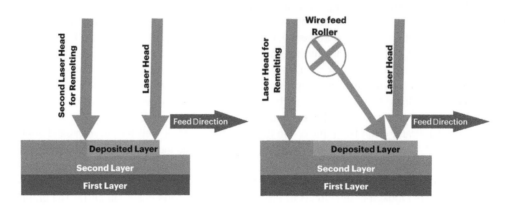

Figure 3.6 (a) Hybrid PBF by laser remelting. (b) Hybrid DED by laser remelting.

following are the advantages of HAM by remelting: (1) improved component density, (2) reduction of residual tensions (Shiomi et al., 2004), (3) changing material characteristics, and (4) improved resistance to fatigue and toughness.

Disadvantages of HAM by remelting include a longer processing time. This is because the energy beam must perform at least twice as many scans. The greater energy needed for processing is another downside.

Yasa et al. reported how laser remelting SLM sections affected exterior surface roughness, microstructure, part density, and hardness (Yasa et al., 2010, 2011a, 2011b). The averaged porosity was reduced by remelting from 0.77% (SLM-only component) to 0.032% when appropriate laser parameters were used. They make use of high scanning speed (100–200 mm/s) and low laser power (85 W). Interestingly, many remelting laser passes will not lower porosity much. When enough energy was supplied, the microstructure gradually developed into something like a lamellar pattern exhibiting finer grain size. This typically resulted in enhanced microhardness (Campanelli et al., 2013).

3.5.5 HAM by Ablation or Erosion

In this process, utilizing an electron beam or a laser, the upper layer of accumulated material is ablated or eroded. Ablation amid build layers is an integrated non-contact approach that has a synergistic effect on the quality and performance of the product. This subtractive method, comparable to HAM by machining, may generate smooth and accurate surfaces by eliminating material. Because the energy source erodes the material, this process may also be utilized to micromachine interlayer features in additive manufacturing.

For selective laser erosion, Yasa et al. investigated HAM on AISI 316 L stainless steel utilizing SLM and a pulsed Q-switched Nd:YAG laser (=1,094 nm) (Shiomi et al., 2004; Yasa et al., 2010). This technique can improve build direction (z-direction) accuracy by lowering layer thickness or rectifying anomalies. They reported that they achieved a 50% decrease in roughness using this process. The schematic diagram of the process is shown in Figure 3.7.

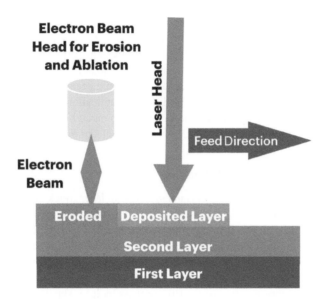

Figure 3.7 Hybrid PBF by erosion or ablation.

3.5.6 HAM by Laser-Assisted Plasma Deposition

Unlike conventional AM procedures, this hybrid AM approach of laser-assisted plasma deposition (LAPD) facilitates plasma deposition with the application of an assisted laser. The plasma deposits the deposited material while the laser adds an additional energy source that is simultaneously applied to the same location to aid the build process and strengthen the printing quality. The laser in LAPD does not help with the building process exactly the same a laser helps with laser-assisted turning cutting. By prolonging tool life or decreasing cutting pressure through work softening, the laser instantly helps the cutting process in laser-assisted turning. The plasma deposition process in LAPD is unaffected by the laser. Instead, the assisting laser adds energy to the plasma production process, improving the part's construction.

The influence of a heat source (laser) on plasma arc deposition was examined by Qian et al. (Qian et al., 2008). This example shows how an aiding energy source can increase both performance of the process and the quality of the product in additive manufacturing. To create greater heat energy, the laser is placed in the plasma arc beam. In plasma arc deposition, the shielding gas absorbed the energy and ionized the gas molecules. The energy density of the plasma arc has been raised, while the diameter of the plasma arc has been reduced, improving composition accuracy. Using the plasma generated by the laser, creating an arc is easy. As more power is used, the depth of the weld puddle increases, improving the microstructure and reducing porosity.

3.5.7 HAM by Peening

Surface treatments for HAM are an area of research that has not received much attention yet. A German industry, MTU Aero Engines, claimed the use of rolling, laser shock peening,

ultrasonic impact treatment, and shot peening to harden specific portions of 3D printed turbine blades in a patent application submitted in 2012 (Bamberg et al., 2012). Additional companies, including United Technologies Corp., General Electric Co., and others, submitted
equivalent patents between 2014 and 2020 (El-Wardany et al., 2020; Wu et al., 2020). When
it comes to implementing surface treatments into additive manufacturing, there is a significant
knowledge gap. Surface treatments, for example, ultrasonic peening and laser shock peening,
are poorly understood. These are crucial for different applications in various sectors such as
defense, aviation, auto industry, and biomedical field, to name but a few. Surface treatments
enable hybrid AM, allowing for unparalleled breakthroughs in material design.

3.5.7.1   HAM by Laser Shock Peening

In this process, the primary AM process is integrated with laser shock peening (LSP) as a secondary process. Like other HAM processes, laser shot peening is a sequential process. As the
chain of cyclical processes, this is also performed layer after layer or perhaps after multiples
layers. After laser peening is used to lase the highest layer, the main process is utilized to generate another layer or layers. The whole process is repeated until the build is completed. This
method enables functionally gradient traits to be used across the produced volume. Figure 3.8
shows the schematic representation of the PBF process by laser shock peening.

In laser shock peening, a workpiece is permanently deformed by the application of shock
waves from intensifying plasma. When a workpiece collides with a pulsed laser having a
range of nanoseconds, plasma is created. To protect the workpiece's surface against heat
damage, an ablative layer is frequently employed. The ablative layers bear the brunt of the
heat stress, while the workpiece is simply subjected to the growing plasma's shock wave.
Here, LSP is basically mechanical in implementation. LSP can also happen without the presence of a protective layer. LSP acts as a thermo-mechanical process, in this case, resulting in
remelted or recast material as well as a shock wave.

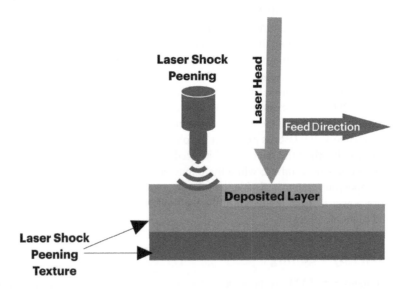

Figure 3.8 PBF by laser shock peening.

It has been observed that in LB-PBF of stainless steel (Kalentics et al., 2017; Ding et al., 2006), the application of LSP after every layer is necessary to achieve lower residual stress in compression. Likewise, to achieve higher residual stress in compression, LSP can be applied after numerous layers.

3.5.7.2   HAM by Ultrasonic Peening

This process makes use of an electromechanical transducer for the layer-by-layer or multiple-layer deployment of ultrasonic energy to a workpiece. Ultrasonic impact treatment is a surface treatment process that can reduce residual stress in compression, alleviate tension, and improve microstructural grain. AM components' fatigue, corrosion, and tribological performance can all be improved by ultrasonic peening. Figure 3.9 shows hybrid DED and PBF by ultrasonic peening. Using ultrasonic peening in HAM, to enhance properties in practically all AM processes is a low-cost, quick, and easy solution.

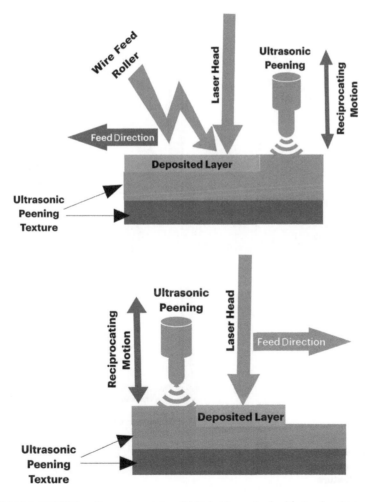

Figure 3.9 (a) Hybrid DED by ultrasonic peening. (b) Hybrid powder bed fusion by ultrasonic penning.

Achuthan et al. studied the use of ultrasonic peening layer-upon-layer to strengthen SLM sections (Gale et al., 2016). The findings demonstrated that ultrasonic peening HAM was effective in enhancing the yield strength of Inconel and stainless steel while also refining the microstructure.

3.5.7.3 Hybrid AM by Shot Peening

Another procedure that incorporates surface treatment is shot peening. Guiding a stochastic stream of beads at extremely high speeds under regulated coverage circumstances enhance the mechanical properties of a near-surface layer. Plastic deformation is caused by the collision of the bead on the surface. As a result, residual compression stress and strain hardening occur. Beads might be made of glass, metal, or ceramic. The utilization of tiny particle shot peening on AlSi10Mg during a PBF process was examined by Sangid et al. (Book et al., 2016). Shot peening throughout the building process provides a variety of benefits, but it also poses additional processing concerns. Shot peening, for example, is a quick and

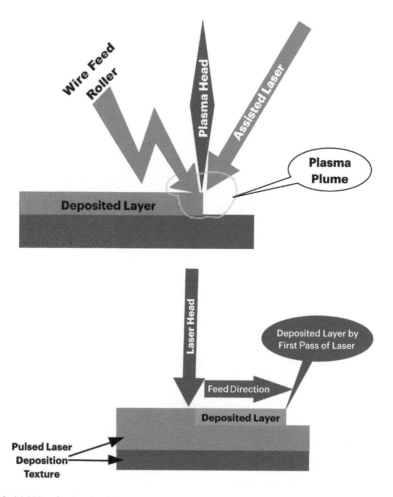

Figure 3.10 (a) Wire feed pulsed laser deposition. (b) PBF by pulsed laser deposition.

low-cost way to strengthen surface integrity. The size of these beads is usually one to three times that of the powders used in AM treatments and hence they require extra sieving from reprocessed powder and chip cutting in HAM equipment. Conventional shot peening is perfect for DED, sheet lamination, and extrusion techniques since the particle size can be substantially bigger and will not conflict with subsequent printing. When a secondary material is utilized as the peening media, due to part contamination difficulties, shot peening in PBF may become more difficult. The AM powder is being used as a replacement for the peening medium. Shot peening can cause close-tolerance parts to be distorted and a pebbly surface finish by shot peening is not appropriate for a number of applications. Fine particle shot peening is used in such cases. Furthermore, the peening medium has limited strength and hardness. Enough contact pressure may not be generated if soft powders are peened by soft materials, for example, aluminum alloy powder if peened by aluminum alloy.

3.5.7.4 HAM by Pulsed Laser Deposition

The use of pulsed lasers is becoming more common in additive manufacturing (Gibson et al., 2021). These lasers with high-powered characteristics are being used onto a substrate to print thin layers of components (Lunney et al., 1995). In this process, a pulsed laser has impinged on the powder. This causes the vaporization of the powder because of instant heating. Vaporization gives rise to plasma plume production. This plume develops a shock wave during printing which causes permanent deformation of the surface (Morgan et al., 2001)

The mechanism of both pulsed laser deposition (PLD) and laser shot peening is almost identical. The only difference is, in a single laser source, PLD coupled with both peening and printing together. PLD allows for favorable compressive residual stresses (Zhou et al., 2003).

3.6 COLD SPRAY+: AN EMERGING HYBRID ADDITIVE MANUFACTURING TECHNOLOGY

A relatively recent solid-state protective layer technology that has acquired sizable consideration is cold spray additive manufacturing (CSAM). This is due to the fact that it is well-suited for the mass fabrication of metal components and metal matrix composite materials. CSAM has notably been proposed as a method for producing/fixing components or attempting to make a welded connection (Li et al., 2019). In comparison to other thermal spraying methods, CSAM functions with hardly any effect of heat on the injection component (base). As a result, the CSAM is widely employed to deposit nanocrystals and amorphous components (temperature-sensitive materials), while also materials such as aluminum, titanium, and copper (non-ferrous oxygen-sensitive materials). Furthermore, CSAM has the opportunity to rapidly create complex self-contained forms that are superior to other AM methods.

Historically, CSAM has been used in manufacturing, repair, readjustment, and welding. They include various industrial metals, for example, aluminum, tin, copper, diamond-reinforced metal matrix composites named DMMC, Ni-Al composites, highly reactive metals such as titanium, and tantalum, and high entropy alloys (HEA). The alloy also can be deposited on the substrate. It was also used in the manufacture of components, gears, crankshafts, and sputtering targets. It has also been used to repair and restore various defective components in order to recover lost size and usage. The process has also been used to connect non-ferrous metals. Nonetheless, the CSAM deposits are defective and have little or no ductility/plasticity due to the mechanism of process deposition, which may be distinguished by intense permanent deformation of the individual particles during layer formation. This flaw limits its wide range of industrial applications.

To exhibit the good strength of CSAMed deposits, specific hybrid technology could be used. Laser machining, post-spray heat treatment, friction stir welding, shot peening, and hot hydrostatic pressing are a few examples. The deposit's cohesive strength, toughness, and bond strength between both the deposit and the substrate can be advanced.

3.7 FUTURE CHALLENGES IN HAM

Hybrid additive manufacturing has significant benefits over traditional and AM processes. Researchers are working on many areas related to hybrid manufacturing to make it more palatable in the market. Their key objectives are to (i) increase the number of appropriate materials, (ii) enhance the finished product's surface properties, (iii) create one-of-a-kind design tools to boost efficiency, (iv) lower the costs of machinery (Dilberoglu et al., 2021), and (v) bring down the cost of production.

HAM has the following research prospects:

1. *In-process measurement:* Because the total process is so complicated in comparison to traditional manufacturing activities, there is a chance for a lot of unexpected things to happen. Broken tools, surface flaws, and unexplained changes in process parameters all impair the overall quality of the parts throughout manufacturing. For registering such occurrences, in-process measurement instruments (such as measurement probes, and non-tactile sensors like high-resolution cameras) could be implemented into hybrid production systems. By dynamically sending the corresponding information (on dimensions) to the CNC units and conducting corrective actions on a frequent basis, the accuracy of the produced items might be greatly enhanced.

2. *Advanced process planning:* Hybrid process planning has lately been used solely for basic topologies due to its higher complexity in contrast to traditional methodologies. Engineers, planners, and designers may benefit from more research on how to effectively manage subsequent/simultaneous additive and subtractive processes. Indeed, manual procedures that rely on the operator's skill have largely dominated modern hybrid process design practices. As a result, industrial uptake of hybrid AM may be hampered by a scarcity of manufacturing talent. Hence, sophisticated software tools are necessary to address difficulties such as successfully arranging the tool path while also regulating the AM process parameters.

3. *Materials:* Since the material must be suitable for both machining and AM processes, the number of materials that may be processed using hybrid manufacturing machinery is restricted. As a result, new materials for hybrid AM processes should be evaluated in the near future, as should new types of process combinations capable of handling such materials. Furthermore, recent advancements in several AM technologies have hastened the use of multiple materials, resulting in a significant expansion of the design domain. Secondary processing on such composite layers (i.e., simultaneous machining of numerous dissimilar materials) may raise key challenges in the poorly developed selection of various parameters, for example, feed rate, cutting speed, tool material, and tool geometry.

4. ***Tool Design:*** With properly controlled subtractive equipment, such as new forming tools instead of machining tools, the cost of HAM can be decreased. Dynamic heads can be employed in the HAM to enhance the capabilities of the process. In the case of material extrusion operations, the tip cross section of the extruder can be adjusted continuously rather than the fixed cross section during the build. This increases the product complexity while decreasing the time of fabrication.

5. ***Management of chips:*** Chip (dust/swarf) creation is unavoidable because machining is an extremely important element of hybrid manufacturing. These chips should be handled properly for high-quality parts. Otherwise, AM activities will be disrupted. New methods must be devised for chip management as well as for environmental concerns inside machine tools in the case of HAM.

3.8   SUMMARY

This chapter describes why hybrid additive manufacturing is significant, the distinction between HAM and hybrid manufacturing, the different HAM processes (assisted or mixed), hybrid materials, and hybrid equipment. In addition, the chapter discussed the new HAM technology Cold Spray+. The last section provides an overview of potential issues in hybrid additive manufacturing.

HAM is a new area of research and innovation in the manufacturing industry. Academia and industry are increasingly interested in learning more about and exploring the advantages of hybrid capabilities. It is expected that more and more manufacturers will soon enter the HAM industry. This will lead to a decline in the cost of high-performance products. This will further increase the demand for economic as well as sustainability evaluation instruments that are new or updated to analyze the life cycle of components designed and manufactured by HAM. Sensing technology may also be essential withinside the assessment and reputation of hybrid AM technology.

REFERENCES

Akula, Sreenathbabu, and K. P. Karunakaran. "Hybrid adaptive layer manufacturing: An Intelligent art of direct metal rapid tooling process." *Robotics and Computer-Integrated Manufacturing* 22, no. 2 (2006): 113–123.

Bamberg, J., T. Hess, R. Hessert, and W. Satzger. "Verfahren zum herstellen, reparieren oder austauschen eines bauteils mit verfestigen mittels druckbeaufschlagung." *Mtu Aero Engines Gmbh, Munich, Germany, Patent No. WO 2012152259* (2012): A1. https://patentscope.wipo.int/search/de/detail.jsf?docId=WO2012152259

Book, Todd A., and Michael D. Sangid. "Evaluation of select surface processing techniques for in situ application during the additive manufacturing build process." *JOM* 68, no. 7 (2016): 1780–1792.

Brecher, Christian, Michael Emonts, Chris-Jörg Rosen, and Jan-Patrick Hermani. "Laser-assisted milling of advanced materials." *Physics Procedia* 12 (2011): 599–606.

Brehl, D. E., and T. A. Dow. "Review of vibration-assisted machining." *Precision Engineering* 32, no. 3 (2008): 153–172.

Campanelli, S. L., G. Casalino, N. Contuzzi, and A. D. Ludovico. "Taguchi optimization of the surface finish obtained by laser ablation on selective laser molten steel parts." *Procedia CIRP* 12 (2013): 462–467.

Choi, Doo-Sun, S. H. Lee, B. S. Shin, K. H. Whang, Y. A. Song, S. H. Park, and H. S. Jee. "Development of a direct metal freeform fabrication technique using CO2 laser welding and milling technology." *Journal of Materials Processing Technology* 113, no. 1–3 (2001): 273–279.

CIRP. "CIRP—The internal academy for production engineering." (2011). https://www.cirp.net/

Colegrove, Paul A., Jack Donoghue, Filomeno Martina, Jianglong Gu, Philip Prangnell, and Jan Hönnige. "Application of bulk deformation methods for microstructural and material property improvement and residual stress and distortion control in additively manufactured components." *Scripta Materialia* 135 (2017): 111–118.

Dilberoglu, Ugur M., Bahar Gharehpapagh, Ulas Yaman, and Melik Dolen. "Current trends and research opportunities in hybrid additive manufacturing." *The International Journal of Advanced Manufacturing Technology* 113, no. 3 (2021): 623–648.

Ding, Kan, and Lin Ye. *Laser Shock Peening: Performance and Process Simulation*. Woodhead Publishing, Cambridge, England, 2006.

El-Wardany, Tahany Ibrahim, Matthew E. Lynch, Daniel V. Viens, and Robert A. Grelotti. "Turbine disk fabrication with in situ material property variation." *U.S. Patent* 10, 710, 161 (July 14, 2020).

Fessler, J. R., R. Merz, A. H. Nickel, Fritz B. Prinz, and L. E. Weiss. "Laser deposition of metals for shape deposition manufacturing." In *1996 International Solid Freeform Fabrication Symposium*, 1996. http://hdl.handle.net/2152/69928

Flynn, Joseph M., Alborz Shokrani, Stephen T. Newman, and Vimal Dhokia. "Hybrid additive and subtractive machine tools–Research and industrial developments." *International Journal of Machine Tools and Manufacture* 101 (2016): 79–101.

Francis, Romy, Joseph Newkirk, and Frank Liou. "Investigation of forged-like microstructure produced by a hybrid manufacturing process." *Rapid Prototyping Journal* 22, no. 4 (2016): 717–726.

Friel, Ross J., and Russell A. Harris. "Ultrasonic additive manufacturing–a hybrid production process for novel functional products." *Procedia Cirp* 6 (2013): 35–40.

Gale, J., A. Achuthan, and A. U. Don. "Material property enhancement in additive manufactured materials using an ultrasonic peening technique." In *Solid Freeform Fabrication Symposium (SFF)*. Austin, TX, August 2016, pp. 8–10.

Gibson, Ian, David Rosen, Brent Stucker, and Mahyar Khorasani. "Hybrid additive manufacturing." In *Additive Manufacturing Technologies*. Springer, Cham, 2021, pp. 347–366.

Hur, Junghoon, Kunwoo Lee, and Jongwon Kim. "Hybrid rapid prototyping system using machining and deposition." *Computer-Aided Design* 34, no. 10 (2002): 741–754.

Ichimura, M., Y. Urushisaki, K. Amaya, S. Chappell, M. Honnami, M. Mochizuki, and U. Chung. "Medical implant manufacture using the hybrid metal laser sintering with machining process." In *15th International Conference on Precision Engineering (ICPE)*. Kanazawa, Japan, July 2014, pp. 22–25.

Jeng, Jeng-Ywan, and Ming-Ching Lin. "Mold fabrication and modification using hybrid processes of selective laser cladding and milling." *Journal of Materials Processing Technology* 110, no. 1 (2001): 98–103.

Kalentics, Nikola, Eric Boillat, Patrice Peyre, Cyril Gorny, Christoph Kenel, Christian Leinenbach, Jamasp Jhabvala, and Roland E. Logé. "3D Laser Shock Peening–A new method for the 3D control of residual stresses in selective laser melting." *Materials & Design* 130 (2017): 350–356.

Kalentics, Nikola, Roland Logé, and Eric Boillat. "Method and device for implementing laser shock peening or warm laser shock peening during selective laser melting." *U.S. Patent* 10, 596, 661 (March 24, 2020).

Karunakaran, K. P., Vishal Pushpa, Sreenath Babu Akula, and S. Suryakumar. "Techno-economic analysis of hybrid layered manufacturing." *International Journal of Intelligent Systems Technologies and Applications* 4, no. 1–2 (2008): 161–176.

Koshy, Philip, V. K. Jain, and G. K. Lal. "Grinding of cemented carbide with electrical spark assistance." *Journal of Materials Processing Technology* 72, no. 1 (1997): 61–68.

Lauwers, Bert, Fritz Klocke, Andreas Klink, A. Erman Tekkaya, Reimund Neugebauer, and Don Mcintosh. "Hybrid processes in manufacturing." *CIRP Annals* 63, no. 2 (2014): 561–583.

Lee, Choon-Man, Wan-Sik Woo, Dong-Hyeon Kim, Won-Jung Oh, and Nam-Seok Oh. "Laser-assisted hybrid processes: A review." *International Journal of Precision Engineering and Manufacturing* 17, no. 2 (2016): 257–267.

Li, Wenya, Congcong Cao, Guoqing Wang, Feifan Wang, Yaxin Xu, and Xiawei Yang. "'Cold spray+' as a new hybrid additive manufacturing technology: A literature review." *Science and Technology of Welding and Joining* 24, no. 5 (2019): 420–445.

Lunney, James G. "Pulsed laser deposition of metal and metal multilayer films." *Applied Surface Science* 86, no. 1–4 (1995): 79–85.

Matsubara, K., Y. Miyahara, Z. Horita, and T. G. Langdon. "Developing superplasticity in a magnesium alloy through a combination of extrusion and ECAP." *Acta Materialia* 51, no. 11 (2003): 3073–3084.

Merz, Robert, F. B. Prinz, K. Ramaswami, M. Terk, and L. E. Weiss. "Shape deposition manufacturing." In *1994 International Solid Freeform Fabrication Symposium*, 1994.

Morgan, R., C. J. Sutcliffe, and W. O'Neill. "Experimental investigation of nanosecond pulsed Nd: YAG laser re-melted pre-placed powder beds." *Rapid Prototyping Journal* 7, no. 3 (2001): 159–172.

Nau, B., A. Roderburg, and F. Klocke. "Ramp-up of hybrid manufacturing technologies." *CIRP Journal of Manufacturing Science and Technology* 4, no. 3 (2011): 313–316.

Ngo, Tuan D., Alireza Kashani, Gabriele Imbalzano, Kate T. Q. Nguyen, and David Hui. "Additive manufacturing (3D printing): A review of materials, methods, applications and challenges." *Composites Part B: Engineering* 143 (2018): 172–196.

Palanivel, S., P. Nelaturu, B. Glass, and R. S. Mishra. "Friction stir additive manufacturing for high structural performance through microstructural control in an Mg based WE43 alloy." *Materials & Design (1980–2015)* 65 (2015): 934–952.

Qian, Ying-Ping, Ju-Hua Huang, Hai-Ou Zhang, and Gui-Lan Wang. "Direct rapid high-temperature alloy prototyping by hybrid plasma-laser technology." *Journal of Materials Processing Technology* 208, no. 1–3 (2008): 99–104.

Sealy, M. P., G. Madireddy, C. Li, and Y. B. Guo. "Finite element modeling of hybrid additive manufacturing by laser shock peening." In *2016 International Solid Freeform Fabrication Symposium*. Austin, TX: University of Texas at Austin, 2016.

Sealy, Michael P., Gurucharan Madireddy, Robert E. Williams, Prahalada Rao, and Maziar Toursangsaraki. "Hybrid processes in additive manufacturing." *Journal of Manufacturing Science and Engineering* 140, no. 6 (2018).

Shiomi, M., K. Osakada, K. Nakamura, T. Yamashita, and F. Abe. "Residual stress within metallic model made by selective laser melting process." *Cirp Annals* 53, no. 1 (2004): 195–198.

Wang, Z. Y., and K. P. Rajurkar. "Cryogenic machining of hard-to-cut materials." *Wear* 239, no. 2 (2000): 168–175.

Wang, Zhiyuan, Renwei Liu, Todd Sparks, Heng Liu, and Frank Liou. "Stereo vision based hybrid manufacturing process for precision metal parts." *Precision Engineering* 42 (2015): 1–5.

Wu, Sung-Yueh, Chen Yang, Wensyang Hsu, and Liwei Lin. "3D-printed microelectronics for integrated circuitry and passive wireless sensors." *Microsystems & Nanoengineering* 1, no. 1 (2015): 1–9.

Wu, Zhiwei, Yanmin Li, David Henry Abbott, Xiaobin Chen, Thomas Froats Broderick, Judson Sloan Marte, Andrew Philip Woodfield, and Eric Allen Ott. "Method for manufacturing objects using powder products." *General Electric Co U.S. Patent* 10, 780, 501 (September 22, 2020).

Xinhong, Xiong, Zhang Haiou, Wang Guilan, and Wang Guoxian. "Hybrid plasma deposition and milling for an aeroengine double helix integral impeller made of superalloy." *Robotics and Computer-Integrated Manufacturing* 26, no. 4 (2010): 291–295.

Yasa, Evren, Jan Deckers, and Jean-Pierre Kruth. "The investigation of the influence of laser re-melting on density, surface quality and microstructure of selective laser melting parts." *Rapid Prototyping Journal* 17, no. 5 (2011a): 312–327.

Yasa, Evren, and J.-P. Kruth. "Investigation of laser and process parameters for Selective Laser Erosion." *Precision Engineering* 34, no. 1 (2010): 101–112. Yasa, Evren, J.-P. Kruth, and Jan Deckers.

"Manufacturing by combining selective laser melting and selective laser erosion/laser re-melting." *CIRP Annals* 60, no. 1 (2011b): 263–266.

Zhou, Xiangman, Haiou Zhang, Guilan Wang, Xingwang Bai, Youheng Fu, and Jingyi Zhao. "Simulation of microstructure evolution during hybrid deposition and micro-rolling process." *Journal of Materials Science* 51, no. 14 (2016): 6735–6749.

Zhou, Y. C., Z. Y. Yang, and X. J. Zheng. "Residual stress in PZT thin films prepared by pulsed laser deposition." *Surface and Coatings Technology* 162, no. 2–3 (2003): 202–211.

Chapter 4

Progress in Metal Additive Manufacturing

Fabrication Techniques, Materials Used, and Potential Applications

Adil Wazeer, Apurba Das, and Shrikant Vidya

4.1 INTRODUCTION

Additive manufacturing is considered as a new methodology having budded prospective to revolutionize traditional production. When compared to traditional methods, AM procedures assist in reducing the requirement for tooling and give opportunities for innovation strategy modification (Fidan et al., 2019; Joshi et al., 2015). Hideo Kodama, who pioneered on the use of UV lamps to harden polymers and build solid things in the early 1980s, achieved the initial advances in AM (Kodama, 1981). Employment of AM with metals, the most regularly used engineering materials, began only in the early 1990s with the introduction of binder jetting by Ely Sachs and collaborators (Sachs et al., 1993). AM technologies may utilize extensive material choice, notably ceramic, polymers, as well as metals. Within these materials, metallic materials are getting prominence among academics and enterprises. Additionally, metal AM may give certain sustainability benefits such as reduced waste, improved quality, lower pollutant discharges, and the ability to produce components on requirement (Huang et al., 2016; Rejeski et al., 2018). Though metal AM offers such advantages, its application in industry are still confined up to limited industries, including aerospace (Najmon et al., 2019), dentistry (Galante et al., 2019), and construction (Buchanan et al., 2019). Furthermore, researchers have mainly concentrated on the exploration of a limited number of metal AM methods.

Pan et al. (2018) examined the mechanical characteristics of metallic materials, namely Ti alloy, and presented arc welding AM techniques. Ahn (2016) evaluated several direct metal AM methods and delivered restrictions associated with their use. Ziaee et al. (2019) discussed binder jetting procedures, materials, and contemporary technology. Deb Roy et al. (Wu et al., 2019), for example, presented microstructures, flaws, as well as mechanical characteristics of materials that were metallic produced/printed. Yakout and team (Yakout et al., 2018) investigated impact of critical printing settings upon mechanical characteristics in addition to morphology of a variety of metal materials, including Ti-, Al-, and Ni-based alloys. Deb Roy and team (DebRoy et al., 2018) also studied metal AM materials, focusing on precious metals, refractory alloys, and compositionally graded alloys.

From this, it is clear that metal AM is extensively available from the standpoint of material creation by many investigators; nevertheless, the usage and applicability of diverse metal AM need additional exploration. Undoubtedly, an outline is necessary that presents all major current conventional metal AM technologies as well as a more in-depth investigation of industry uses are mandatory.

This chapter aims at providing a complete analysis of present status of metal AM, presenting major popular techniques, and reviewing industrial applications through many industry

DOI: 10.1201/9781003270027-4

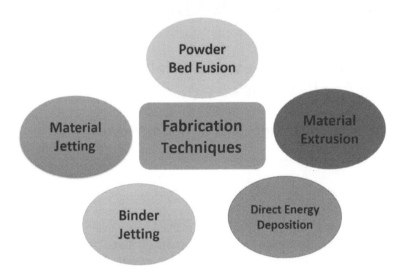

Figure 4.1 Different fabrication techniques that are employed in the metal AM process.

divisions (Figure 4.1). Furthermore, exploration required in extensive usage of metal AM for a number of industrial segments, as well as investigation and expansion initiatives, was extracted and reviewed in this chapter.

4.2 FABRICATION TECHNIQUES

4.2.1 Powder Bed Fusion

Powder bed fusion methods are employed in most metal AM systems. Selective laser melting, direct metal laser melting, direct metal laser sintering, and electron beam melting (EBM) are popular metal PBF methods. Generally, there is not much variation between the forms of these laser techniques produced by diverse industries. Furthermore, there is minimal change in materials utilized by the various machines. DMLS and SLM procedures are the only ones that are utilized to manufacture metallic parts (Patterson et al., 2017). SLM, also known as laser metal fusion, is mostly employed in metals like Al, but direct metal laser melting is typically intended for alloys of metals like Al and Ti (Delgado et al., 2012). The direct metal laser melting method exists equivalent to direct metal laser sintering procedure; apart from that, beams generate ultrafine layers of metal and deliver homogeneous melted pool in DMLM method (Berlier et al., 2018). This method has several benefits over DMLS, including enhanced quality of surface and reduced porosity levels. Rather than a laser in SLM printers, a higher intensity electron beam is employed in EBM technique to fuse metal powders together. Electron beam additive manufacturing (EBAM) or selective electron beam melting (SEBM) are other names for this technology (Calignano et al., 2017). Because components may be arranged in the build volume, EBM procedure results in increased productivity compared to SLM methods; yet, larger degrees of deformation and residual stresses exist in the AM components owing to superior energy density and speedy thermal cycles (Choi et al., 2019).

4.2.2 Material Extrusion

In 1989, Stratasys Ltd. delivered a patent for the fused deposition modeling (FDM) technique, which employs thermoplastic polymers (Wohlers et al., 2017). Metal-filled filaments that could be produced with FFF machines are among the most current advancements in metal AM sector. Metal powder stays incorporated inside a typical filament of PLA and thus the powder proportion might fluctuate. The prints generated with these filaments are impure metal pieces, characterized as metal particles in polymer matrix, thus they are fragile and must be handled with care (Riecker et al., 2016). The items produced by these methods are not metallic; instead, they have a metal composition. It must be emphasized that when the metal particle concentration rises, the filaments' tensile strength decreases; nevertheless, the filaments' thermal conductivity improves, making them appropriate for constructing circuits and electromagnetic devices (Hwang et al., 2015). A new advancement in the metal AM sector is bound powder extrusion (BPE). The filament in this technique is made up from plastic binder plus minute metal powder which was ejected from nozzle. Required item can be produced in layers in a manner similar to standard FFF printing, although the method must be carried out via two extra post-processing procedures—washing and sintering, resulting in finished higher density component of metal.

4.2.3 Direct Energy Deposition

From an energy standpoint, DED can be divided into two groups: cold spray and the thermal energy (Dass et al., 2019). Cold spray is a technique for depositing material as small particles onto a surface having enough kinetic energy for forming dense and thick layer (Villafuerte, 2014). Another type of DED apparatus uses an electron beam, laser beam, arc or plasma to generate thermal energy. This group melts the feedstock material, which could be powder or wire, and delivers it onto the build platform in stages (Moridi et al., 2020). This additive manufacturing group employs computerized welding techniques for generating through superior rates of deposition, however with less resolution (Cunningham et al., 2018).

A DED technique involves melting the substrate and then applying metal powder to the molten layer. Blown powder technology (BPT) (Chiumenti et al., 2017) is the name of this technique. Such methods, unlike many other powder AM procedures, need not employ a powder bed to deposit the metallic material; instead, the powder and laser are concentrated onto the surface for metallic material deposition. Furthermore, metal powder is poured in a molten pool of metal and a heating basis is acted by concentrated beam of laser, allowing for high-precision manufacturing of near-net form components (Onuike et al., 2018). Additionally, the BPT comprises numerous methodologies according to the application or condition of the technique (Zenou et al., 2018; Ghosal et al., 2018).

Another DED technique is LC, which involves employing laser heat for depositing fine layer of preplaced materials onto a surface stirred via monitored mechanism (Liu et al., 2017). LC consists of two feeding schemes, according to Toyserkani and team (Yue et al., 2007): a one-step method which comprises paste feeding, injection of powder, and feeding of wire, whereas a two-step methodology termed replacement powder. The cladding materials were deposited on surface where shallow melted pool is generated simultaneously in the one-step scheme; in the two-step route, the cladding material is accumulated onto surface as powder and therefore beam of laser is ablated.

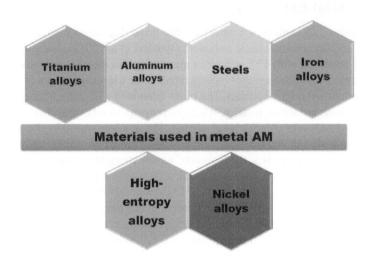

Figure 4.2 Various materials that are used in metal AM techniques for various applications in numerous fields.

4.2.4   Binder Jetting

Binder jetting method prints the necessary metal components in layers using powder of metal in addition to liquid binder. Droplets of binder combine powdered components inside and between sliced layers in this procedure (Wheat et al., 2018). When assessed with metal (PBF) powder bed fusion, the BJ method needs no backing assemblies and generates far better and precise items.

4.2.5   Material Jetting

Photopolymers are used as the material in material jetting technique, and UV light is used to selectively treat the printed layer. Nanoparticle jetting (NPJ) is among the material jetting methods, which allows metal objects to be manufactured using composite resins impregnated with metallic nanoparticles like stainless steel (SS). The MJ technique allows for the creation of materials of higher resolution that is among the primary disadvantages in traditional additive manufacturing methods (Figure 4.2).

4.3   MATERIALS USED

4.3.1   Titanium Alloys

Ti alloys are considered among widely researched metallic materials in AM field. Having superior specific strength plus resistance to fracture, outstanding formability, and great resistance to corrosion as well as fatigue, and strong biocompatibility, titanium alloys are extensively employed in biomedical and aerospace applications (Bandyopadhyay et al., 2019a). Many investigations have found that alternative AM techniques, like DED and PBF, could be used to treat Ti alloys (Bandyopadhyay et al., 2019a; Gangireddy et al., 2019; Nicoletto et al., 2017). Owing to fast solidification in AM processing, Ti alloys show columnar grains in its morphology. Such microstructure type is frequent for AM-generated components; it also tends to develop in layers alongside the build track. Anisotropic qualities in AM-treated components are caused by the shape of columnar grains, according to studies (Bandyopadhyay et al., 2019b; Kontis et al., 2019). Researchers discovered an acicular martensitic α' phase in AM-generated Ti6Al4V samples (Liu et al., 2019), inclined toward improvement of strength of samples of Ti-6Al-4V while decreasing their ductility. Post–heat treatments on Ti6Al4V components are commonly used to enhance ductility through decaying α' phase into phases of α and β (Bandyopadhyay et al., 2019a; Gangireddy et al., 2019). When compared to bulk Ti, the mechanical characteristics of additively manufactured porous and lattice Ti alloy structures showed exceptional energy captivation capabilities and resistance to impact (Yang et al., 2019). Because 3D generated Ti constructs were previously authorized through the US Food and Drug Administration (FDA), additional design amendment using AM of titanium alloys might result in considerable advantages for medical implants.

4.3.2   Aluminum Alloys

Owing to weak absorption of laser and lower welding ability, present aluminum alloy that could be simply additively created are still restricted. Al-Si-Mg and Eutectic Al-Si alloys are extensively utilized Al alloys in AM (for instance; AlSi10Mg and Al12Si). Silicon is included in such alloys that boost laser absorptivity (Zhang et al., 2019). DED-produced Al12Si material exhibited a finer morphology having eutectic silicon incorporated into matrix of Al that improves thermal characteristics, according to research (Espana et al., 2011). Due to the small grain microstructure created by AM process, AlSi10Mg treated using powder bed fusion displayed outstanding resistance to cavitation erosion compared with the identical material generated via casting in another work (Girelli et al., 2018).

4.3.3   Steels

AM has been used to treat a variety of steels, comprising martensitic, austenitic, duplex, maraging, as well as precipitation hardened steels. AM-manufactured steels have diverse microstructures and phases of precipitation than traditionally manufactured steels, which can contribute to mechanical property variations (Yan et al., 2019). Owing to fast solidification and non-equilibrium circumstances, the morphologies of AM-treated steels include finer and crystallographically textured characteristics (Yan et al., 2019). Heat treatment is commonly used to achieve required characteristics in AM-manufactured steels. According to studies, SS 316L treated using the laser-based PBF technology produces entirely austenitic and columnar grains having grain size of around 1 m that is substantially finer than traditionally manufactured SS 316L (Yu et al., 2017; Yasa et al., 2011). Furthermore, experiments have shown that DED-treated SS 316L may produce both austenitic and ferritic phases. Micro-segregation following solidification in DED process resulted in elevation of chromium and molybdenum, both being ferrite stabilizers (Ziętala et al., 2016). Whereas intercellular area of the PBF-produced SS 316L had an abundance for Cr and Mo, the level of ferritic phase stabilizer was insufficient to stable the ferritic phase area. Moreover, investigators find that PBF-produced materials like austenitic stainless steel improve strength despite sacrificing ductility (Yan et al., 2019). It is now feasible to control and supervise the cooling rate using modern AM equipment by modifying processing conditions to produce tailored mechanical characteristics for various steels.

4.3.4 Iron and Nickel Alloys

In powder-based AM techniques, SS 316L (UNS S31603) is used much often. The density and, as a result, the mechanical characteristics of the manufactured components are affected by powder particle sizes starting with the feedstock (Spierings et al., 2009; Spierings et al., 2011). Several studies found that point distance, exposure duration, scanning speed, thickness of layer, and orientation of the building all have a momentous influence upon superiority of components generated in SLM of SS 316L. To achieve a satisfactory surface quality and mechanical characteristics, these factors should be regulated during the manufacturing process (Spierings et al., 2012; Cherry et al., 2015). Special alloys, including nickel-based alloys, may now be fabricated using AM technique. According to certain investigations, AM Inconel 718 (UNS N07718) components had microscopic fractures that might impact mechanical qualities in all dimensions, particularly in the building direction (Kistler, 2015). The phase transition as well as the production of columnar dendrites throughout melting procedure is responsible for these fissures (Gong et al., 2015). Several researches, on either hand, have shown how to choose process variables for manufacturing dense components from Invar 36 (UNS K93600) (Qiu et al., 2016).

4.3.5 Other Alloys

For its unusual features, Ni-based alloys and high-entropy alloys had received a lot of interest. Such alloys are typically utilized in applications requiring harsh circumstances, like extreme temperatures, corrosive environments, and complicated loading situations. Due to their reduced machinability, nickel-based alloys are notably complicated to handle. Traditional techniques of processing nickel-based alloys, like powder metallurgy (PM) and casting, cannot create components with complex geometries on a part-by-part approach. When dealing with HEAs, similar concerns arise. Furthermore, traditionally produced HEAs exist limited considering microstructure refinement besides mechanical characteristics (Ho et al., 2018; Guan et al., 2019). Recent findings suggest that metal AM techniques may be utilized to manufacture Ni-based alloys plus HEAs, overcoming the drawbacks of traditional approaches. Furthermore, the materials like TiN particle reinforced CoCrFeNiMn HEA synthesized via SLM exhibited finer and almost equiaxial grains, with a strength of above 1.0 GPa (Li et al., 2020).

4.3.6 Multi-metal Components

The uses of AM methods to manufacture multi-metal components have piqued the interest of researchers. Titanium, aluminum, stainless steel, copper, and nickel are most often utilized metals in MMAM. Using an LENS system, Heer and team created compositionally graded magnetic non-magnetic bimetallic structures. By immediately migrating from non-magnetic austenitic stainless steel 316 (SS316) to magnetic ferritic stainless steel 430 (SS430) in a single structure, a graded magnetic functionality was achieved (Heer et al., 2018). Yusuf and co-researchers studied the interfacial region of laser powder bed fusion (L-PBF)–fabricated multi-material 316L stainless steel/Inconel 718 (316L SS/IN718) (Yusuf et al., 2021). Electron backscattered diffraction and scanning electron microscopy investigations, porosity study, plus Vickers hardness test were used to describe the interfacial area. In comparison with FZ and 316L SS areas, IN718 zone has superior hardness. Tan used laser powder bed

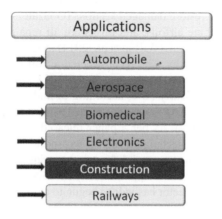

Figure 4.3 Various applications of metal AM in different sectors.

fusion for two-type iron-based multi-materials, AISI 304 stainless steel (SS304) and AISI 1045 carbon steel, to examine the influence of material elemental composition on interfacial bond strength (CS 45). Flexural tests and static tensile test, as well as dynamic tension fatigue analysis, were used to assess mechanical behavior. The results indicate that interfacial bond strength is greater than the substrate materials' fracture strength (Tan et al., 2021). Demir et al. presented their work on multi-material SLM system and also illustrated the application in the fabrication of Fe/Al-12Si multi-material systems. Given the limited suitability and dispersibility of the two materials, the Fe/Al-12Si layers produced huge fractures, but the resulting Fe/Al-12Si layers had a superior hardness (Demir et al., 2017). Singh et al. used a reinforced FDM design in investment casting route to generate an Al/Al$_2$O$_3$ composite as a FGM in an investigation (Singh et al., 2016). Tan et al., for example, used SLM to create a W-Cu (tungsten–copper) FGM. Interfacial flaws, bonding strength, and microstructure were all studied in relation to the laser constraint. Owing to the intrinsic features of materials, anomalous pores and fractures were found to be the most common flaws at the contact (Tan et al., 2019) (Figure 4.3).

4.4   POTENTIAL APPLICATIONS

4.4.1   Aerospace

Metal AM is suited for aerospace industries because of its capacity to perform quick tooling and repair, create freeform and complicated shapes, and combine components (Singamneni et al., 2019). Furthermore, aircraft equipment is frequently composed of challenging materials like superalloys with exceptional resistance to corrosion (Kellner, 2017). Metal brackets and bleed pipes for Airbus airplanes are manufactured utilizing EOS and Concept Laser equipment. Furthermore, in conjunction with Arconic (Najmon et al., 2019), this business is creating massive AM airframe parts. This company's creation of improved A320 parts and reworking titanium portions of the A350 (Wohlers et al., 2017) is yet another move toward utilization of metal AM. Since past few years, aviation business has paid close attention to DED technology. For example, LMD method was utilized to fabricate turbomachinery impeller blades along with restore damaged sections (González-Barrio et al., 2020) Engine combustion chambers, turbine airfoils, and blisks are all repaired using DED technology (Najmon et al., 2019). Wilson et al. (Wilson et al., 2014) used LENS technique to fix defects in turbine airfoils that were faulty. The reconstructed airfoil replicated the original shape with 0.03 mm of average precision, lowering carbon trace by around 45% as well as increasing overall power savings to nearly 36%, according to findings. Fuel nozzles for General Electric (GE) Aviation's new LEAP engine are printed using Arcam Metal printers and Concept Laser (Russell et al., 2019). As per a paper released by Raja and team (Raja et al., 2006), additive manufacturing techniques aid in aerospace field by certain means: cycle time has been condensed down to around 70%, non-recurring expenditures are cut down to nearly 45%, and weight is reduced to approximately 35%. As can be seen from the foregoing discussions, the aerospace sector has indeed begun to enjoy the benefits of adopting AM technology.

4.4.2 Biomedical

Metal AM techniques are very well to manufacture end-use goods in the healthcare and dentistry sectors. Metal additive manufacturing allows such businesses to create unique replicas that are suited to the demands of patients and get access to additional materials. Metal PBF and Arcam's EBM technology have been used by LimaCorporate Company in Italy to manufacture titanium-alloy orthopedic devices (Wohlers et al., 2017). Spine surgery has become increasingly prevalent as a result of aging, trauma, or spinal tumors as per this firm (Graves 2019). Following on a CT scan, Harrysson (Harrysson et al., 2015) created integrated implants and produced the unique Ti6Al4V implants using EBM or DMLS methods. Osseus Fusion Systems, Egan Dental Laboratory, OMX Solutions, Endocon GmbH, and Johnson & Johnson Medical Equipment are just a few firms which have employed additive manufacturing to make orthodontic devices. Vilardell et al. (Vilardell et al., 2019) used lattice structures to create a topology-optimized human. The unique structures were manufactured on an SLM EOS M28 printer using Ti6Al4V. The novel layouts can lessen the effect of stress shielding in graft employment, according to the findings. One PBF printer from Aurora Labs is used for manufacturing an orthodontic strut-like shape in a different investigation (Kain et al., 2020). The findings revealed that the novel product's material qualities result in a lower hardness around the gums.

4.4.3 Automotive

Metal additive manufacturing has enabled more adaptable, optimal, and durable designs, as well as lightweight, tougher, and safe goods, quick customization, and lower cycle time and prices. As a result, a growing number of automakers are turning to metal additive manufacturing. Formula Student Germany in 2012 employed EOS Direct Metal Laser Sintering method for creating an upright of lesser-weight that was 35% lighter than the previous cast part (Jensen, 2015). A metal 3D printer was used by BMW to produce a window guide rail in i8 Roadster. This AM method is currently being utilized to produce 100 rails in 24 hours (Tyrrell, 2015). BMW is also introducing a new engine that will be a successor for the S55 engine. This engine combines a number of parts, along with a cylindrical head made with a PBF metal additive manufacturing printer (Anusci, 2019). The 3D printed part on the i8 Roadster is a clamp utilized in soft-top connection that is composed of Al alloy. The part is nearly 44% less in weight but ten times stiffer than that of the injection-molded plastic attachment which was initially used (Bakewell, 2019). Audi, a German automaker, has teamed up with SLM Solution Group to produce bespoke goods, replacement components, and parts that are only needed once in a while (Wang et al., 2019). A water adaptor for the Audi W12 engine, for example, was made additively.

4.4.4 Electronics

Metal AM has gained a lot of interest from researchers and industry in recent years because it has the potential to revolutionize the design and production of unique electrical and communication devices such as circuit boards, resistors, waveguides, couplers, and antennas (Espera et al., 2019). These electrical and communication components are printed and/or patterned from functional materials, namely, metal, polymer, and semiconductor. Metal/inorganic are the most often utilized materials in this business sector because they have superior conductivity and endurance (Espera et al., 2019). Many experts have looked at using AM methods to create antennas with unusual shapes. For example, Goh and team (Goh et al., 2016) created patch antennas utilized in wireless systems of lower power using an inkjet printer. For communication satellites, Robert Hofmann Company generated one lighter weight filter for RF (Wohlers et al., 2017). Another usage of metal AM is in creating high-efficiency metal waveguides. A W-band component was designed by Verploegh et al. (Verploegh et al., 2017), which incorporates a part of straight waveguide of 10 cm plus a coupler of 20 dB. The part is made in maraging Steel using DMLS technology, which gives structural integrity and appropriate conductivity.

4.4.5 Some Other Applications

Associated with longer printing durations, smaller printing volumes, higher initial costs, and environmental concern, the employment of metal AM methods for construction usage is presently restricted; nevertheless, the number of applications in this sector is growing. Metal additive manufacturing allows the construction sector to manufacture complicated structural structures with less weight plus shorter cycle time. The metal AM is now utilized to produce bespoke parts on necessity. Moreover, metal AM is considered advantageous in situations when building elements are damaged or lost, and on-site fabrication utilizing AM could be advantageous because delaying for alternatives may create delays in construction projects and expenses (Camacho et al., 2017).

Mining activities are frequently carried out in distant and high-risk areas. As a result, when system failures occur, supply chains are threatened owing to access restrictions, downtime, and manufacturing issues caused by long lead times for replacement parts. Metal AM could be ideal in certain situations, especially for the production of on-demand spare components (Frandsen et al., 2020). Mold and tool companies are progressively using metal AM. Metal AM is utilized to create tools like mold inserts, fittings, jigs, and gauges in this sector (Kang et al., 2018). A 3D generated construct, for example, having skeleton shape with lattice shell and rib reinforcement elements. Such construct allows for quick as well as consistent cool off. The improved AM mold is 30% higher effective than that of initial plus reduced distortion, residual stresses, as well as casting faults, according to the findings (Kang et al., 2018).

The railway sector has begun to use AM technology for manufacturing and conservation. Alstom, a French international establishment, utilizes additive manufacturing to create lightweight parts (Killen et al., 2018). Dubai's Road Transport Authority's preservation team is making struggles for employing additive manufacturing methods to different metro rail assets, like creating AM components for ticketing. Furthermore, railway firms have difficulty locating spare components that were abandoned or whose makers have gone out of business, and which will be prohibitively costly to replicate utilizing traditional production techniques (Kingsland, 2019).

4.5   CONCLUSIONS

Metal additive manufacturing provides several benefits over conventional manufacturing techniques in terms of creating new designs, complicated structures, and bespoke parts with the least amount of waste. This chapter presented a thorough examination of typical metal additive manufacturing technologies, including recent breakthroughs and current industrial uses.

- The automobile, aviation, biomedical, and dentistry sectors are the dominant sectors in the metal AM industry. Other sectors, such as petroleum, electronics, infrastructure, and railway, have been paying attention to metal AM in recent times as a method to make substantial gains in the development and manufacturing of innovative products.
- The interrelationships between metal AM process conditions, mechanical characteristics, and geometrical precision for any specific metal AM technique are currently unknown. As a result, various theoretical theories will have to be built. Heat and mass transit, residual stress and porosity forecasting, deformation assessment, change of phase, as well as other features may be included in these models.
- Another research gap in metal AM is how to decrease energy consumption and waste by applying appropriate metal AM methods and, as a result, implementing just-in-time manufacturing.
- Furthermore, employing innovative materials and/or procedures, as well as topology optimization approaches, the performance of manufactured metal AM parts may be enhanced.
- Furthermore, more investigation is necessary to describe the most critical characteristics for metal AM development and build a judgment model to assist industries in determining where metal AM will be most appropriate.

Finally, creating thorough, generally acknowledged technical specifications for metal AM is indeed a work in progress in order to ensure metal print quality uniformity.

ACKNOWLEDGMENTS

The authors acknowledge Prof. Amit Karmakar of Jadavpur University, Kolkata, and Dr. Arijit Sinha of Kazi Nazrul University, Asansol, for providing guidance during preparation of this work.

REFERENCES

Ahn, D.G., 2016. Direct metal additive manufacturing processes and their sustainable applications for green technology: A review. *International Journal of Precision Engineering and Manufacturing-Green Technology*, 3(4), pp. 381–395.

Anusci, V., 2019. *BMW's New S58 Engine Features Cylinder Head Made with 3D Printing*. Available online: www.3dprintingmedia.network/bmw-s58-engine-3d-printed-cylinder/

Bakewell, J., 2019. *Customizing Production*. Available online: www.automotivemanufacturingsolutions.com/customisingproduction/31218.article.

Bandyopadhyay, A., Shivaram, A., Mitra, I. and Bose, S., 2019a. Electrically polarized TiO_2 nanotubes on Ti implants to enhance early-stage osseointegration. *Acta Biomaterialia*, 96, pp. 686–693.

Bandyopadhyay, A., Upadhyayula, M., Traxel, K.D. and Onuike, B., 2019b. Influence of deposition orientation on fatigue response of LENS™ processed Ti6Al4V. *Materials Letters*, 255, p. 126541.

Berlier, J., McCann, A., Zhang, L. and Good, B., 2018. Systems and methods for receiving sensor data for an operating additive manufacturing machine and adaptively compressing the sensor data based on process data which controls the operation of the machine. *Google Patents US20180348734A1*, 6 December 2018. Available online: https://patents.justia.com/patent/10635085.

Buchanan, C. and Gardner, L., 2019. Metal 3D printing in construction: A review of methods, research, applications, opportunities and challenges. *Engineering Structures*, 180, pp. 332–348.

Calignano, F., Manfredi, D., Ambrosio, E.P., Biamino, S., Lombardi, M., Atzeni, E., Salmi, A., Minetola, P., Iuliano, L. and Fino, P., 2017. Overview on additive manufacturing technologies. *Proceedings of the IEEE*, 105(4), pp. 593–612.

Camacho, D.D., Clayton, P., O'Brien, W., Ferron, R., Juenger, M., Salamone, S. and Seepersad, C., 2017. Applications of additive manufacturing in the construction industry–a prospective review. In *ISARC. Proceedings of the International Symposium on Automation and Robotics in Construction* (Vol. 34). IAARC Publications.

Cherry, J.A., Davies, H.M., Mehmood, S., Lavery, N.P., Brown, S.G.R. and Sienz, J., 2015. Investigation into the effect of process parameters on microstructural and physical properties of 316L stainless steel parts by selective laser melting. *The International Journal of Advanced Manufacturing Technology*, 76(5), pp. 869–879.

Chiumenti, M., Lin, X., Cervera, M., Lei, W., Zheng, Y. and Huang, W., 2017. Numerical simulation and experimental calibration of additive manufacturing by blown powder technology. Part I: Thermal analysis. *Rapid Prototyping Journal*, 23(2), pp. 448–463.

Choi, Y. and Lee, D.G., 2019. Correlation between surface tension and fatigue properties of Ti-6Al-4V alloy fabricated by EBM additive manufacturing. *Applied Surface Science*, 481, pp. 741–746.

Cunningham, C.R., Flynn, J.M., Shokrani, A., Dhokia, V. and Newman, S.T., 2018. Invited review article: Strategies and processes for high quality wire arc additive manufacturing. *Additive Manufacturing*, 22, pp. 672–686.

Dass, A. and Moridi, A., 2019. State of the art in directed energy deposition: From additive manufacturing to materials design. *Coatings*, 9(7), p. 418.

DebRoy, T., Wei, H.L., Zuback, J.S., Mukherjee, T., Elmer, J.W., Milewski, J.O., Beese, A.M., Wilson-Heid, A.D., De, A. and Zhang, W., 2018. Additive manufacturing of metallic components–process, structure and properties. *Progress in Materials Science*, 92, pp. 112–224.

Delgado, J., Ciurana, J. and Rodríguez, C.A., 2012. Influence of process parameters on part quality and mechanical properties for DMLS and SLM with iron-based materials. *The International Journal of Advanced Manufacturing Technology*, 60(5), pp. 601–610.

Demir, A.G. and Previtali, B., 2017. Multi-material selective laser melting of Fe/Al-12Si components. *Manufacturing Letters*, 11, pp. 8–11.

Espana, F.A., Balla, V.K. and Bandyopadhyay, A., 2011. Laser processing of bulk Al–12Si alloy: Influence of microstructure on thermal properties. *Philosophical Magazine*, 91(4), pp. 574–588.

Espera, A.H., Dizon, J.R.C., Chen, Q. and Advincula, R.C., 2019. 3D-printing and advanced manufacturing for electronics. *Progress in Additive Manufacturing*, 4, pp. 245–267.

Fidan, I., Imeri, A., Gupta, A., Hasanov, S., Nasirov, A., Elliott, A., Alifui-Segbaya, F. and Nanami, N., 2019. The trends and challenges of fiber reinforced additive manufacturing. *The International Journal of Advanced Manufacturing Technology*, 102(5), pp. 1801–1818.

Frandsen, C.S., Nielsen, M.M., Chaudhuri, A., Jayaram, J. and Govindan, K., 2020. In search for classification and selection of spare parts suitable for additive manufacturing: A literature review. *International Journal of Production Research*, 58(4), pp. 970–996.

Galante, R., Figueiredo-Pina, C.G. and Serro, A.P., 2019. Additive manufacturing of ceramics for dental applications: A review. *Dental Materials*, 35(6), pp. 825–846.

Gangireddy, S., Komarasamy, M., Faierson, E.J. and Mishra, R.S., 2019. High strain rate mechanical behavior of Ti-6Al-4V octet lattice structures additively manufactured by selective laser melting (SLM). *Materials Science and Engineering: A*, 745, pp. 231–239.

Ghosal, P., Majumder, M.C. and Chattopadhyay, A., 2018. Study on direct laser metal deposition. *Materials Today: Proceedings*, 5(5), pp. 12509–12518.

Girelli, L., Tocci, M., Montesano, L., Gelfi, M. and Pola, A., 2018. Investigation of cavitation erosion resistance of AlSi10Mg alloy for additive manufacturing. *Wear*, 402, pp. 124–136.

Goh, G.L., Ma, J., Chua, K.L.F., Shweta, A., Yeong, W.Y. and Zhang, Y.P., 2016. Inkjet-printed patch antenna emitter for wireless communication application. *Virtual and Physical Prototyping*, 11(4), pp. 289–294.

Gong, X. and Chou, K., 2015. Microstructures of Inconel 718 by selective laser melting. In *TMS 2015 144th Annual Meeting & Exhibition* (pp. 461–468). Springer, Cham.

González-Barrio, H., Calleja-Ochoa, A., Lamikiz, A. and López de Lacalle, L.N., 2020. Manufacturing processes of integral blade rotors for turbomachinery, processes and new approaches. *Applied Sciences*, 10(9), p. 3063.

Graves, J., 2019. *Nexxt Spine Develops AM Spinal Implants Using MTS Test Systems* (p. 89). Nexxt Spine, Noblesville, IN.

Guan, S., Solberg, K., Wan, D., Berto, F., Welo, T., Yue, T.M. and Chan, K.C., 2019. Formation of fully equiaxed grain microstructure in additively manufactured AlCoCrFeNiTi0. 5 high entropy alloy. *Materials & Design*, 184, p. 108202.

Harrysson, O.L., Marcellin-Little, D.J. and Horn, T.J., 2015. Applications of metal additive manufacturing in veterinary orthopedic surgery. *JOM*, 67(3), pp. 647–654.

Heer, B. and Bandyopadhyay, A., 2018. Compositionally graded magnetic-nonmagnetic bimetallic structure using laser engineered net shaping. *Materials Letters*, 216, pp. 16–19.

Ho, I.T., Chen, Y.T., Yeh, A.C., Chen, C.P. and Jen, K.K., 2018. Microstructure evolution induced by inoculants during the selective laser melting of IN718. *Additive Manufacturing*, 21, pp. 465–471.

Huang, R., Riddle, M., Graziano, D., Warren, J., Das, S., Nimbalkar, S., Cresko, J. and Masanet, E., 2016. Energy and emissions saving potential of additive manufacturing: The case of lightweight aircraft components. *Journal of Cleaner Production*, 135, pp. 1559–1570.

Hwang, S., Reyes, E.I., Moon, K.S., Rumpf, R.C. and Kim, N.S., 2015. Thermo-mechanical characterization of metal/polymer composite filaments and printing parameter study for fused deposition modeling in the 3D printing process. *Journal of Electronic Materials*, 44(3), pp. 771–777.

Jensen, W., 2015. *Automotive: Formula Student Germany—EOS Supports Racing Team by Producing a Topology-Optimized Steering Stub Axle*. Available online: www.eos.info/press/customer_case_studies/rennteam_uni_stuttgart.

Joshi, S.C. and Sheikh, A.A., 2015. 3D printing in aerospace and its long-term sustainability. *Virtual and Physical Prototyping*, 10(4), pp. 175–185.

Kain, M., Nadimpalli, V., Miqueo, A., May, M., Yagüe-Fabra, J., Häfner, B., Pedersen, D., Calaon, M., Tosello, G. and Kain, M., 2020, June. Metal additive manufacturing of multi-material dental strut implants. In *Proceedings of the 20th International Conference of the European Society for Precision Engineering and Nanotechnology (EUSPEN 20), EUSPEN* (pp. 8–12). Geneva, Switzerland.

Kang, J., Shangguan, H., Deng, C., Hu, Y., Yi, J., Wang, X., Zhang, X. and Huang, T., 2018. Additive manufacturing-driven mold design for castings. *Additive Manufacturing*, 22, pp. 472–478.

Kellner, T., 2017. *3D-Printed 'Bionic' Parts Could Revolutionize Aerospace Design*. General Electric Reports: Boston, MA.

Killen, A., Fu, L., Coxon, S. and Napper, R., 2018, December. Exploring the use of additive manufacturing in providing an alternative approach to the design, manufacture and maintenance of interior rail components. In *Proceedings of the 40th Australasian Transport Research Forum (ATRF 2018)* (Vol. 30). Darwin, Australia.

Kingsland, P., 2019. *3D Printing in the Railway Sector with Deutsche Bahn*. Available online: www. railway-technology.com/features/3d-printing-in-the-railway-sector/

Kistler, N.A., 2015. *Characterization of Inconel 718 Fabricated Through Powder Bed Fusion Additive Manufacturing*. B.Sc. Thesis, Materials Science and Engineering, Pennsylvania State University.

Kodama, H., 1981. Automatic method for fabricating a three-dimensional plastic model with photo-hardening polymer. *Review of Scientific Instruments*, 52(11), pp. 1770–1773.

Kontis, P., Chauvet, E., Peng, Z., He, J., da Silva, A.K., Raabe, D., Tassin, C., Blandin, J.J., Abed, S., Dendievel, R. and Gault, B., 2019. Atomic-scale grain boundary engineering to overcome hot-cracking in additively-manufactured superalloys. *Acta Materialia*, 177, pp. 209–221.

Li, B., Zhang, L., Xu, Y., Liu, Z., Qian, B. and Xuan, F., 2020. Selective laser melting of CoCrFeNiMn high entropy alloy powder modified with nano-TiN particles for additive manufacturing and strength enhancement: Process, particle behavior and effects. *Powder Technology*, 360, pp. 509–521.

Liu, J., Yu, H., Chen, C., Weng, F. and Dai, J., 2017. Research and development status of laser cladding on magnesium alloys: A review. *Optics and Lasers in Engineering*, 93, pp. 195–210.

Liu, S. and Shin, Y.C., 2019. Additive manufacturing of Ti6Al4V alloy: A review. *Materials & Design*, 164, p. 107552.

Moridi, A., Stewart, E.J., Wakai, A., Assadi, H., Gartner, F., Guagliano, M., Klassen, T. and Dao, M., 2020. Solid-state additive manufacturing of porous Ti-6Al-4V by supersonic impact. *Applied Materials Today*, 21, p. 100865.

Najmon, J.C., Raeisi, S. and Tovar, A., 2019. Review of additive manufacturing technologies and applications in the aerospace industry. *Additive Manufacturing for the Aerospace Industry*, pp. 7–31.

Nicoletto, G., Maisano, S., Antolotti, M. and Dall'Aglio, F., 2017. Influence of post fabrication heat treatments on the fatigue behavior of Ti-6Al-4V produced by selective laser melting. *Procedia Structural Integrity*, 7, pp. 133–140.

Onuike, B., Heer, B. and Bandyopadhyay, A., 2018. Additive manufacturing of Inconel 718—Copper alloy bimetallic structure using laser engineered net shaping (LENS™). *Additive Manufacturing*, 21, pp. 133–140.

Pan, Z., Ding, D., Wu, B., Cuiuri, D., Li, H. and Norrish, J., 2018. Arc welding processes for additive manufacturing: A review. *Transactions on Intelligent Welding Manufacturing*, pp. 3–24.

Patterson, A.E., Messimer, S.L. and Farrington, P.A., 2017. Overhanging features and the SLM/DMLS residual stresses problem: Review and future research need. *Technologies*, 5(2), p. 15.

Qiu, C., Adkins, N.J. and Attallah, M.M., 2016. Selective laser melting of Invar 36: Microstructure and properties. *Acta Materialia*, 103, pp. 382–395.

Raja, V., Zhang, S., Garside, J., Ryall, C. and Wimpenny, D., 2006. Rapid and cost-effective manufacturing of high-integrity aerospace components. *The International Journal of Advanced Manufacturing Technology*, 27(7), pp. 759–773.

Rejeski, D., Zhao, F. and Huang, Y., 2018. Research needs and recommendations on environmental implications of additive manufacturing. *Additive Manufacturing*, 19, pp. 21–28.

Riecker, S., Clouse, J., Studnitzky, T., Andersen, O. and Kieback, B., 2016. Fused deposition modeling-opportunities for cheap metal AM. *World PM2016-AM-Deposition Technologies*. Available online: https://www.ifam.fraunhofer.de/content/dam/ifam/en/documents/dd/WorldPM2016/ Riecker_Fused%20Deposition%20Modeling%20%E2%80%93%20Opportunities%20For%20 Cheap%20Metal%20AM.pdf

Russell, R., Wells, D., Waller, J., Poorganji, B., Ott, E., Nakagawa, T., Sandoval, H., Shamsaei, N. and Seifi, M., 2019. Qualification and certification of metal additive manufactured hardware for aerospace applications. *Additive Manufacturing for the Aerospace Industry*, pp. 33–66.

Sachs, E.M., Haggerty, J.S., Cima, M.J., Williams, P.A., 1993. *U.S. Patent No. 5,204*. U.S. Patent and Trademark Office, Washington, DC, p. 55.

Singamneni, S., Yifan, L.V., Hewitt, A., Chalk, R., Thomas, W. and Jordison, D., 2019. Additive manufacturing for the aircraft industry: A review. *Journal of Aeronautics & Aerospace Engineering*, 8(1), pp. 351–371.

Singh, S. and Singh, R., 2016. Development of functionally graded material by fused deposition modelling assisted investment casting. *Journal of Manufacturing Processes*, 24, pp. 38–45.

Spierings, A.B., Herres, N. and Levy, G., 2011. Influence of the particle size distribution on surface quality and mechanical properties in AM steel parts. *Rapid Prototyping Journal*, 17(3), pp. 195–202.

Spierings, A.B. and Levy, G., 2009, September. Comparison of density of stainless steel 316L parts produced with selective laser melting using different powder grades. In *2009 International Solid Freeform Fabrication Symposium*. University of Texas at Austin.

Spierings, A.B., Wegener, K. and Levy, G., 2012, August. Designing material properties locally with additive manufacturing technology SLM. In *2012 International Solid Freeform Fabrication Symposium*. University of Texas at Austin.

Tan, C., Wang, D., Ma, W. and Zhou, K., 2021. Ultra-strong bond interface in additively manufactured iron-based multi-materials. *Materials Science and Engineering: A*, 802, p. 140642.

Tan, C., Zhou, K. and Kuang, T., 2019. Selective laser melting of tungsten-copper functionally graded material. *Materials Letters*, 237, pp. 328–331.

Tyrrell, M. 2015. *Use of 3D Printed Components at BMW Jumps 42% Annually*. Available online: www.pesmedia.com/3dprinting-components-bmw-group/

Verploegh, S., Coffey, M., Grossman, E. and Popović, Z., 2017. Properties of 50–110-GHz waveguide components fabricated by metal additive manufacturing. *IEEE Transactions on Microwave Theory and Techniques*, 65(12), pp. 5144–5153.

Vilardell, A.M., Takezawa, A., Du Plessis, A., Takata, N., Krakhmalev, P., Kobashi, M., Yadroitsava, I. and Yadroitsev, I., 2019. Topology optimization and characterization of Ti6Al4V ELI cellular lattice structures by laser powder bed fusion for biomedical applications. *Materials Science and Engineering: A*, 766, p. 138330.

Villafuerte, J., 2014. Considering cold spray for additive manufacturing. *Advanced Materials & Processes*, 50, pp. 50–52.

Wang, Y., Lin, Y., Zhong, R.Y. and Xu, X., 2019. IoT-enabled cloud-based additive manufacturing platform to support rapid product development. *International Journal of Production Research*, 57(12), pp. 3975–3991.

Wheat, E., Vlasea, M., Hinebaugh, J. and Metcalfe, C., 2018. Sinter structure analysis of titanium structures fabricated via binder jetting additive manufacturing. *Materials & Design*, 156, pp. 167–183.

Wilson, J.M., Piya, C., Shin, Y.C., Zhao, F. and Ramani, K., 2014. Remanufacturing of turbine blades by laser direct deposition with its energy and environmental impact analysis. *Journal of Cleaner Production*, 80, pp. 170–178.

Wohlers, T., Campbell, I., Diegel, O., Huff, R. and Kowen, J., 2017. *3D Printing and Additive Manufacturing State of the Industry: Annual Worldwide Progress Report*. Lund University: Lund, Sweden.

Wu, Q., Mukherjee, T., Liu, C., Lu, J. and DebRoy, T., 2019. Residual stresses and distortion in the patterned printing of titanium and nickel alloys. *Additive Manufacturing*, 29, p. 100808.

Yakout, M., Elbestawi, M.A. and Veldhuis, S.C., 2018. A review of metal additive manufacturing technologies. *Solid State Phenomena*, 278, pp. 1–14.

Yan, J.J., Chen, M.T., Quach, W.M., Yan, M. and Young, B., 2019. Mechanical properties and cross-sectional behavior of additively manufactured high strength steel tubular sections. *Thin-Walled Structures*, 144, p. 106158.

Yang, K., Wang, J., Jia, L., Yang, G., Tang, H. and Li, Y., 2019. Additive manufacturing of Ti-6Al-4V lattice structures with high structural integrity under large compressive deformation. *Journal of Materials Science & Technology*, 35(2), pp. 303–308.

Yasa, E. and Kruth, J.P., 2011. Microstructural investigation of Selective Laser Melting 316L stainless steel parts exposed to laser re-melting. *Procedia Engineering*, 19, pp. 389–395.

Yu, H., Yang, J., Yin, J., Wang, Z. and Zeng, X., 2017. Comparison on mechanical anisotropies of selective laser melted Ti-6Al-4V alloy and 304 stainless steel. *Materials Science and Engineering: A*, 695, pp. 92–100.

Yue, T.M., Su, Y.P. and Yang, H.O., 2007. Laser cladding of Zr65Al7. 5Ni10Cu17. 5 amorphous alloy on magnesium. *Materials Letters*, 61(1), pp. 209–212.

Yusuf, S.M., Zhao, X., Yang, S. and Gao, N., 2021. Interfacial characterisation of multi-material 316L stainless steel/Inconel 718 fabricated by laser powder bed fusion. *Materials Letters*, 284, p. 128928.

Zenou, M. and Grainger, L., 2018 Additive manufacturing of metallic materials. In *Additive Manufacturing* (pp. 53–103). Elsevier, Amsterdam, The Netherlands.

Zhang, Y. and Bandyopadhyay, A., 2019. Direct fabrication of bimetallic Ti6Al4V+ Al12Si structures via additive manufacturing. *Additive Manufacturing*, 29, p. 100783.

Ziaee, M. and Crane, N.B., 2019. Binder jetting: A review of process, materials, and methods. *Additive Manufacturing*, 28, pp. 781–801.

Ziętala, M., Durejko, T., Polański, M., Kunce, I., Płociński, T., Zieliński, W., Łazińska, M., Stępniowski, W., Czujko, T., Kurzydłowski, K.J. and Bojar, Z., 2016. The microstructure, mechanical properties and corrosion resistance of 316 L stainless steel fabricated using laser engineered net shaping. *Materials Science and Engineering: A*, 677, pp. 1–10.

Chapter 5

Selection of Advanced 3D Printing Technologies for Piston Casting by Analytical Hierarchy Process

Rupshree Ozah and Manapuram Muralidhar

5.1 INTRODUCTION

India is fourth largest automobile manufacturer in the world. Automotive companies must choose key manufacturing strategies to succeed and prosper in today's global marketplace, which is marked by globalization, the consumer value demands, widening regulatory enforcement, the global economic crisis, and intense competitive pressure. As a result, the manufacturing process chosen is critical in meeting consumer demand. Any expanding economy relies heavily on the automotive industry. The Indian automobile sector is expanding at a rapid pace, with annual sales exceeding one million automobiles. The automobile industry grows at a rate of 10–15% per year, whereas the automobile component sector grows at a rate of 28% per year (Sinha, 2012).

Piston is the critical component for an automobile engine. The piston transfers the thermal energy of the fuel into mechanical energy in an internal combustion engine. Pistons are put to a lot of mechanical and heat stress. Extreme pressure cycles in the combustion chamber with peak pressures up to 200 bar and massive forces of inertia induced by extremely rapid acceleration during the reciprocating motion of the piston cause mechanical strains on the piston. There exist various commonly used conventional manufacturing technologies for piston.

The 3D printing is an emerging technology which has various advantages such as shorter cycle times, lower manufacturing costs, and smaller part weight. It was developed in 1986 by Charles W Hall. Since then, large numbers of processes have been evolved. Additive manufacturing, also known as three-dimensional printing (3DP), has received increasing attention around the world as a result of the widespread use of digital technology in a variety of applications. It is a process of constructing real objects from digital data piece by piece, line by line, surface by surface, or layer by layer, utilizing 3D digital modeling tools and programs. The 3D printing is a type of printing that allows you to make or copy freestanding complex constructions in one piece. A 3D printer's size can range from modest to enormous, up to 4,000 × 2,000 × 1,000 mm building space for larger object (Voxeljet, 2022). The 3D printers have several advantages, including the capacity to manufacture high-quality workpieces with accuracy, convenience, speedy design and printing, cheap cost, ease of use, and the potential to adapt to a variety of applications (Ha et al., 2018). For prioritizing the different additive manufacturing processes, analytical hierarchy process (AHP) is employed.

AHP is known as a transparent selection process from very long. Saaty (1990) has discussed briefly about the analytic hierarchy process—a multi-criteria decision-making approach in which variables are organized in a hierarchic framework. The AHP employs three principles

DOI: 10.1201/9781003270027-5

to address problems: decomposition, contrast, and synthesis of priorities (Saaty, 1991). The decomposition theory is used to construct a hierarchy, in which each level's elements are independent of those in subsequent levels, working down from the target at the top to the specific options at the bottom. The AHP has been employed in a variety of situations, including school admissions, military personnel promotions and employment choices, sports, and determining the best resettlement location for the Turkish city of Adapazari, which was damaged by an earthquake (Saaty, 2008). This study uses the AHP approach to prioritize different additive manufacturing processes for piston manufacturing depending on various attributes. The priority weights (PW) of different attributes were calculated for six 3D printing processes.

5.2 THREE-DIMENSIONAL PRINTING TECHNIQUES FOR PISTON CASTING

Since the advent of the 3D printing process in 1986, several processes have been evolved. In recent years, due to increasing demand for both product complexity and multi-functionality, many new materials, such as nanomaterials, functional materials, biomaterials, smart materials, and even fast drying concrete, have been explored for 3D printability and used as feed materials for printing real application parts. The 3D printing applications include automobile (15%), aerospace (18%), industry (19%), electronics (13%), medical (11%), academic (8%), military and others (16%), which is shown in Figure 5.1.

Fused deposition modelling (FDM), selective laser sintering (SLS), multi-jet modeling (MJM), or stereolithography are all 3D printing technologies that uses polymer materials. There are several methods for creating 3D models directly from metallic powders such as SLS, EBM (electron beam melting), DMLS (direct metal laser sintering), and SLM (selective laser melting).

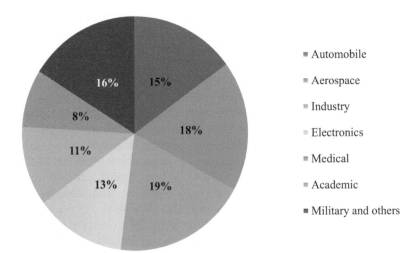

Figure 5.1 Pie chart for 3D printing applications.

Stereolithography (SLA) is a layer-by-layer 3D printing technology that employs photochemical processes to create models, prototypes, patterns, and manufacturing parts. Many different types of prototypes can be created using stereolithography, such as medical models, computer hardware, and prototypes for products in development. Stereolithography has the advantage of producing almost any design but is also expensive. It has very smooth surface finish and average build speed.

Inkjet printing (IJP) technology, also known as binder jetting, is another AM technique that deposits thin layers of material. Both the construction material and the support material are sprayed in separate jets at the same time while being stored in melted liquid states in reservoirs. Individual jetting heads spew droplets as they move in an X–Y orientation when liquids are supplied into them. One of the system's most impressive features is its capacity to generate incredibly fine resolutions and surface finishes, which are basically similar to CNC machines.

Fused deposition modeling (FDM) is one of most important 3D printing technologies in which parts are produced layer by layer using thermoplastic filaments. The filaments generally have a circular cross section with 1.75-mm diameters. In FDM, plastic filaments are unwound from the coil and heated to a semi-liquid state that supplies through an extrusion nozzle. The most commonly used materials are polylactic acid (PLA), nylon, polyethylene terephthalate glycol (PETG), and ABS (acrylonitrile butadiene styrene) (Yadav et al., 2019). The main advantages of this technology include good variety of available materials, easy material change, low maintenance cost, quick production of thin parts, no toxic materials, very compact size, and low temperature operation. The main disadvantages are rough surface, slow process, and redistricted dimensions.

A 3D model of piston is created in a commercial computer-aided design package called AUTOCAD. A 3D model can also be created by scanning a real object via 3D scanner. Based on the data of 3D model, printing can be done. The model is being converted to an STL file format (Standard Tessellation Language) for loading into the slicing software Ultimaker Cura. Slicing software converts model into series of thin layers. Just after the part loading, the printer is set up to print the required data. The required parameters in the Ultimaker Cura setting section are chosen at this stage of sample fabrication. The three-dimensional model is sliced into thin layers and creates the tool path after giving the configuring parameters. The G-code is generated and the file sent to the FDM machine via SD Card. The 3D printer follows the G-code instructions to put down successive layers of liquid, powder, or sheet material to build a model from a series of cross sections of a model.

The nozzle head material is heated up to 1,800 °C which turns the flow on and off. Typically, the stepper motors are employed to move the extrusion head in the Z-direction and adjust the flow according to the requirements. The head may move in both horizontal and vertical directions and the mechanism is controlled by a computer-aided manufacturing (CAM) software program operating on a microcontroller. The steps required for printing any object is shown in Figure 5.2.

A prototype piston has been printed in FDM using PLA, which is shown in Figure 5.3. During FDM printing, the filament material is heated at a certain temperature through the nozzle where it passes through. The nozzle is heated to a temperature (known as nozzle temperature) until the filament melts. The procedure is controlled by a mechanism to control the flow of melted filament. The printing wall thickness of the produced specimen is considered as 0.8 mm while printing speed was 50 mm/sec. Nozzle temperature depends on various aspects such as type of material or printing speed. The nozzle diameter is 0.40 mm.

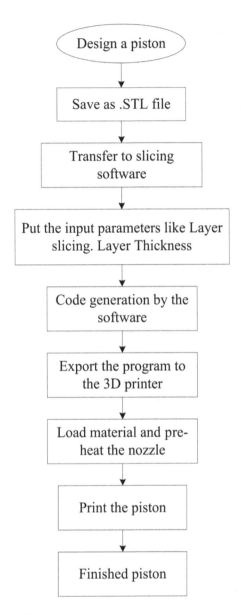

Figure 5.2 The steps involved for printing a piston.

The printing speed (mm/sec) is the one at which the extruder is moved while printing. It measures the amount of manufactured over a given period of time. Getting a shorter printing time required higher value of printing speed. Infill density plays a very important role for printing stronger object. It is the quantity of material used within the printing object. Infill density is the percentage of infill volume with filament material. For printing piston, infill density percentage is considered as 60%.

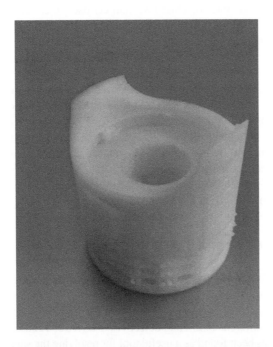

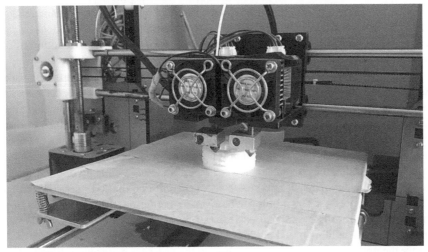

Figure 5.3 A piston constructed by FDM.

In 1991, the first commercialized Laminated Object Manufacturing (LOM) system was delivered. Helisys of Torrance, California, developed LOM (Ramya et al., 2016). A feed mechanism feeds a sheet over a build platform, a heated roller provides pressure to glue the sheet to the layer beneath, and a laser cuts the shape of the part in each sheet layer which are the system's major components.

Initially, selective laser sintering (SLS) was largely used to create prototypes for audiovisual assistance and fit-to-form testing using polymers and nylon. Metal alloys were eventually used to create functioning prototypes. Its goal was to create SLS technology that would allow metal components to be produced directly from powders without the need of a polymer binder or a specially built metal powder.

A novel rapid manufacturing technique called direct metal laser sintering (DMLS) utilizes metallic powders for raw materials. A laser beam sinters metallic powder in layers as the whole part is sliced into several layers. The feature of DMLS allows it to fabricate complex functional parts with high levels of precision, which has led to its use in a variety of industries, including automobile, aerospace, and medicine. Among the most significant concerns with the DMLS process is the build time. Even parts of small–moderate size may require more than 6 hours of processing. Part accuracy should be kept as high as feasible, although reducing construction time might lead to decreased surface accuracy, and vice versa.

5.3 ANALYTICAL HIERARCHY PROCESS

Multi-criteria decision-making strategies are used in the analytic hierarchy process (AHP) approach. The AHP has been widely used in decision-making situations requiring numerous criteria in multilevel systems (Liu et al., 2005). The AHP can be very useful in involving several decision-makers with different conflicting objectives to arrive at a consensus (Tam et al., 2001). The AHP approach has been found as a useful tool for resolving the supplier selection problem and selecting the best supplier combination (Tahriri et al., 2008). Many excellent research works on AHP include numerical extensions of AHP as well as its applications in various disciplines, such as planning, selecting best options, resource allocations, settling dispute, optimization, and so on (Varagas, 1990). The AHP helps to capture both subjective and objective components of a decision by reducing complex judgments to a series of pairwise comparisons and then synthesizing the results. In addition, the AHP includes a beneficial technique for assessing the consistency of the decision-maker's evaluations, thereby eliminating decision-making bias.

The AHP assesses a set of evaluation criteria as well as a set of alternative options from which to select the best option. It is important to remember that because some of the criteria are mutually contradictory, the ideal option is not always the one that optimizes each individual criterion, but rather the one that achieves the best trade-off among all of them. According to the decision-maker's pairwise comparisons of the criteria, the AHP creates a weight for each assessment criterion. The more weight given to a criterion, the more essential it is.

The AHP then awards a score to each option based on the decision-maker's pairwise comparisons of the options based on that criterion for a set criterion. The greater the score, the better the option's performance in relation to the criterion in question. Finally, the AHP combines the criteria weights and option scores to produce a global score for each option and respective ranking as a result. A particular option's overall score is a weighted average of the scores it received across all categories.

5.3.1 Selection of 3D Printing Process by Analytical Hierarchy Process for Piston Manufacturing

AHP is a decision aid that can provide the decision-maker (DM) with relevant information to assist the DM in choosing the "best" alternative or to rank a set of alternatives it allows the decision-makers to check the quality of the results in the comparison matrix.

A stepwise procedure is described here:

- List out various alternative processes such as SLS (selective laser sintering), DMLS (direct metal laser sintering), fused deposition modeling (FDM), laminated object manufacturing (LOM), inkjet printing (IJP), and stereolithography (SLA).
- Identify the various attributes and prioritize them.
- Prepare the attribute matrix A_{10*10}.
- Assign weights 1–9 for each attribute indicating the level of importance from equal to moderate, strong, very strong, to extreme level by 1, 3, 5, 7, and 9, respectively, and their reciprocals in the pairwise comparison matrix in column 1. Cells A11, A22, . . . A1010 will be unity.
- Develop a hierarchy of the problem based on the overall goal, factors, criteria, and decision alternatives.
- Assign each alternative a weight based on its contribution to each decision criterion. This is carried out through a pairwise comparison of the alternatives based on the decision criterion.
- Add the values in each column of the matrix.
- Generate a normalized pairwise comparison matrix, and divide each component in the column by its column totals.
- Calculate the average of the items in each row of the normalized pairwise comparison matrix, which provides an estimate of relative priorities and results into priority vector.
- Compute the overall priority for various attributes (A1–A10) denoted by different processes by

$$\sum PW_i \times PPW_i$$

- Prepare the final composite matrix for attributes A1 to A10 and all processes and compute the overall composite weights.

Six additive manufacturing processes and scale of comparison are presented in Tables 5.1 and 5.2, respectively, according to the stepwise procedure discussed earlier. For priority weights for six 3D printing processes, 9-point scale is considered. Ten different attributes have been taken into measure in this case, as shown in Table 5.3. The relative relevance of the criterion is determined using AHP. The relative relevance of one criterion over another was calculated using pairwise comparisons. To calculate the AHP's final priority weight values, a total of ten pairwise comparisons were performed, as indicated in Table 5.4. The result of pairwise comparison shows that the dimension accuracy is the most preferred criterion and cost issues is the least preferred criterion.

Table 5.1 3D Printing Process Description

	Additive Manufacturing Process	Abbreviation
I	Direct metal laser sintering	DMLS
2	Selective laser sintering	SLS
3	Laminated object manufacturing	LOM
4	Fused deposition modeling	FDM
5	Inkjet printing	IJP
6	Stereolithography	SLA

Table 5.2 Scale for Pairwise Comparison

Degree of Importance	Definition
I	Equally preferred
3	Moderately preferred
5	Strongly preferred
7	Very strongly preferred
9	Extremely preferred
2, 4, 6, 8	Intermediate preferences of one judgment over another

Table 5.3 Description of Attributes for 3D Printing Processes

Attribute	Attribute Name	Attribute Description
AI	Dimensional accuracy	Thickness, diameter, height of the piston
A2	Build area	Printing area of machine
A3	Layer thickness	Thickness of each printing layer
A4	Infill pattern and infill percentage	Amount of infill material for each printing, pattern of printing
A5	Improved mechanical properties	Tensile strength, hardness, elongation
A6	Surface finish	Surface finish should improve
A7	Cycle time	Time required to produce per product
A8	Ease of manufacturing	Easy to manufacture
A9	Process safety	Safety of the process
AI0	Printing cost	Expenses of the process

Table 5.4 Pairwise Comparison of Various Attributes

Attribute	A I	A2	A3	A4	A5	A6	A7	A8	A9	AI0	Priority Weight
AI	I	2	2	3	4	5	6	7	8	8	0.2639
A2	1/2	I	2	2	4	4	5	6	6	7	0.1995
A3	1/2	1/2	I	2	2	3	3	4	5	5	0.1365
A4	1/3	1/2	1/2	I	2	3	4	5	5	6	0.1235
A5	1/4	1/4	1/2	1/2	I	2	2	3	3	4	0.0753
A6	1/5	1/4	1/3	1/3	1/2	I	3	4	4	5	0.0731
A7	1/6	1/5	1/3	1/4	1/2	1/3	I	3	3	4	0.0509
A8	1/7	1/6	1/4	1/5	1/3	1/4	1/3	I	2	3	0.0328
A9	1/8	1/6	1/5	1/5	1/3	1/4	1/3	1/2	I	2	0.0252
AI0	1/8	1/7	1/5	1/6	1/4	1/5	1/4	1/3	1/2	I	0.0192

Pairwise comparison of ten attributes and priority weights were computed for piston casting shown in Table 5.5. Dimension accuracy plays a major role when it comes to manufacture an object and DMLS is found as higher followed by SLS, LOM, FDM, IJP, and SLA. Final composite priority weights are computed and presented in Table 5.6. It shows the relative scale of measurement.

Table 5.5 Pairwise Comparison of Different Additive Manufacturing Processes for Piston Casting

Attribute 1: Pairwise Comparison of Dimensional Accuracy

Process	DMLS	SLS	LOM	FDM	IJP	SLA	Priority weight
DMLS	1	2	3	4	4	5	0.3103
SLS	1/2	1	2	3	4	4	0.2091
LOM	1/3	1/2	1	2	2	3	0.1231
FDM	1/4	1/3	1/2	1	3	4	0.1068
IJP	1/4	1/4	1/2	1/3	1	3	0.0680
SLA	1/5	1/4	1/3	1/4	1/3	1	0.0399

Attribute 2: Pairwise Comparison of Build Area

Process	DMLS	SLS	LOM	FDM	IJP	SLA	Priority weight
DMLS	1	2	3	4	4	5	0.3005
SLS	1/2	1	3	4	4	5	0.2368
LOM	1/3	1/3	1	2	3	4	0.1278
FDM	1/4	1/4	1/2	1	3	4	0.0990
IJP	1/4	1/4	1/3	1/3	1	2	0.0558
SLA	1/5	1/5	1/4	1/4	1/2	1	0.0373

Attribute 3: Pairwise Comparison of Layer Thickness

Process	DMLS	SLS	LOM	FDM	IJP	SLA	Priority weight
DMLS	1	2	3	3	4	5	0.3031
SLS	1/2	1	2	3	3	4	0.2053
LOM	1/3	1/2	1	2	3	3	0.1381
FDM	1/3	1/3	1/2	1	2	3	0.0975
IJP	1/4	1/3	1/3	1/2	1	3	0.0716
SLA	1/5	1/4	1/3	1/3	1/3	1	0.0416

Attribute 4: Pairwise Comparison of Variety of Materials Used for Piston Casting

Process	DMLS	SLS	LOM	FDM	IJP	SLA	Priority weight
DMLS	1	2	3	4	4	6	0.3165
SLS	1/2	1	2	3	4	5	0.2148
LOM	1/3	1/2	1	2	2	3	0.1223
FDM	1/4	1/3	1/2	1	3	4	0.1054
IJP	1/4	1/4	1/2	1/3	1	2	0.0603
SLA	1/6	1/5	1/3	1/4	1/2	1	0.0378

Attribute 5: Pairwise Comparison of Improved Mechanical Properties

Process	DMLS	SLS	LOM	FDM	IJP	SLA	Priority weight
DMLS	1	2	3	3	4	5	0.2999
SLS	1/2	1	2	2	3	4	0.1881
LOM	1/3	1/2	1	2	4	5	0.1593
FDM	1/3	1/2	1/2	1	2	4	0.1063
IJP	1/4	1/3	1/4	1/2	1	3	0.0668
SLA	1/5	1/4	1/5	1/4	1/3	1	0.0367

Attribute 6: Pairwise Comparison of Surface Finish of the Piston

Process	DMLS	SLS	LOM	FDM	IJP	SLA	Priority weight
DMLS	1	2	3	4	5	5	0.3161
SLS	1/2	1	2	3	4	5	0.2116
LOM	1/3	1/2	1	2	3	4	0.1363
FDM	1/4	1/3	1/2	1	3	3	0.0956
IJP	1/5	1/4	1/3	1/3	1	3	0.0602
SLA	1/5	1/5	1/4	1/3	1/3	1	0.0374

Attribute 7: Pairwise Comparison of Cycle Time								Attribute 8: Pairwise Comparison of Ease of Manufacturing							
Process	DMLS	SLS	LO M	FDM	IJP	SLA	Priority weight	Process	DMLS	SLS	LOM	FD M	IJP	SLA	Priority weight
DMLS	I	2	3	4	5	7	0.3296	DMLS	I	3	3	4	5	7	0.3497
SLS	1/2	I	2	3	4	5	0.2116	SLS	1/3	I	2	3	4	4	0.1914
LOM	1/3	½	I	2	2	3	0.1207	LOM	1/3	1/2	I	2	2	3	0.1198
FDM	1/4	1/3	1/2	I	3	4	0.1027	FDM	1/4	1/3	1/2	I	3	4	0.1028
IJP	1/5	¼	1/2	1/3	I	2	0.0566	IJP	1/5	1/4	1/2	1/3	I	2	0.0564
SLA	1/7	1/5	1/3	1/4	1/2	I	0.0361	SLA	1/7	1/4	1/3	1/4	1/2	I	0.0370

Attribute 9: Pairwise Comparison of Process Safety								Attribute 10: Pairwise Comparison of Printing Cost							
Process	DMLS	SLS	LOM	FDM	IJP	SLA	Priority weight	Process	DMLS	SLS	LOM	FDM	IJP	SLA	Priority weight
DMLS	I	2	2	4	5	6	0.3104	DMLS	I	1/2	1/3	1/4	1/4	1/5	0.0394
SLS	1/2	I	2	3	4	4	0.2125	SLS	2	I	1/2	1/3	1/4	1/4	0.0580
LOM	1/2	1/2	I	2	2	3	0.1341	LOM	3	2	I	1/3	1/4	1/4	0.0809
FDM	1/4	1/3	1/2	I	3	3	0.0995	FDM	4	3	3	I	1/4	1/5	0.1260
IJP	1/5	1/4	1/2	1/3	I	2	0.0592	IJP	4	4	4	4	I	1/3	0.2162
SLA	1/6	1/4	1/3	1/3	1/2	I	0.0415	SLA	5	4	4	5	3	I	0.3366

Table 5.6 Final Composite Rating of Processes for Piston Casting

S. No.	Attribute Name	Priority Weight	DMLS	SLS	LOM	FDM	IJP	SLA
I	Dimensional accuracy	0.2639	0.3103	0.2091	0.1231	0.1068	0.068	0.0399
2	Build area	0.1995	0.3005	0.2368	0.1278	0.099	0.0558	0.0373
3	Layer thickness	0.1365	0.3031	0.2053	0.1381	0.0975	0.0716	0.0416
4	Variety of materials	0.1235	0.3165	0.2148	0.1223	0.1054	0.0603	0.0378
5	Improved mechanical properties	0.0753	0.2999	0.1881	0.1593	0.1063	0.0668	0.0367
6	Surface finish	0.0731	0.3161	0.2116	0.1363	0.0956	0.0602	0.0374
7	Cycle time	0.0509	0.3296	0.2116	0.1207	0.1027	0.0566	0.0361
8	Ease of manufacturing	0.0328	0.3497	0.1914	0.1198	0.1028	0.0564	0.037
9	Process safety	0.0252	0.3104	0.2125	0.1341	0.0995	0.0592	0.0415
10	Printing cost	0.0192	0.0394	0.058	0.0809	0.126	0.2162	0.3366
	Total composite weight ($\Sigma PW_i \times PPW_i$)		**0.304814**	**0.210123**	**0.128901**	**0.102779**	**0.066102**	**0.044375**

5.4 DISCUSSION AND CONCLUSIONS

In this work, six 3D printing processes considering ten attributes for manufacturing of the automotive piston using AHP has been considered. The results of analyzing factors are deemed important for 3D printer selection problem using the analytic hierarchy process. According to the result of this study, dimensional precision, printing area, and layer thickness were the most important subfactors influencing the printing machine selection. The 3D printing allows idea for the rapid development. The ability to print a concept the same day it is designed reduces a development process from months to a few days, allowing businesses to keep one step ahead of the competition. Process safety of printing a 3D model includes mechanical hazards and emissions in the form of outgassing and ultrafine particles produced during the filament melting and extrusion process. Safety for DMLS is found higher based on the comparison criterion and weight allocation.

Among six different 3D printing machines, DMLS is found to be have highest composite priority weight of 0.3048 and that can produce a 3DP product with high accuracy, followed with surface finish, part smoothness, reasonable cost price, and build time. Total priority composite weight shows the lowest value, that is, 0.044375 for SLA. The 3D printers are used to manufacture the whole product at once rather than the conventional casting process which requires machining and post-processing afterward. It is useful for selection of 3D printing manufacturing processes for the shop floor engineers. Pairwise comparison of six additive manufacturing processes is done for ten attributes for piston casting. The dimension accuracy for DMLS is found 0.3103 followed by SLS (0.2091), LOM (0.1231), FDM (0.1068), IJP (0.068), and SLA (0.0399). A variety of materials may be used in different 3D manufacturing processes. Choosing the right machine could give the product in right quantity and at right cost on the right timeline.

ACKNOWLEDGMENT

The authors gratefully acknowledge the support extended by North-Eastern Regional Institute of Science and Technology (NERIST), Itanagar, Arunachal Pradesh, India, in the present work.

REFERENCES

Ha, Sangho, Kasin Ransikarbum, Hweeyoung Han, Daeil Kwon, Hyeonnam Kim, and Namhun Kim. "A dimensional compensation algorithm for vertical bending deformation of 3D printed parts in selective laser sintering." *Rapid Prototyping Journal* 24, no. 6 (October 15, 2018): 955–963. https://doi.org/10.1108/RPJ-12-2016-0202.

Liu, Fuh, and Hui Hai. "The voting analytic hierarchy process method for selecting supplier." *International Journal of Production Economics* 97, no. 3 (2005): 308–317.

Official website of Voxeljet. www.voxeljet.com/industrial-3d-printer/serial-production/vx4000/

Ramya, A., and Sai Leela Vanapalli. "3D printing technologies in various applications." *International Journal of Mechanical Engineering and Technology* 7, no. 3 (June 2016): 396–409.

Saaty, Thomas L. "How to make a decision: The analytic hierarchy process." *European Journal of Operational Research* 48, no. 1 (September 1990): 9–26. https://doi.org/10.1016/0377-2217(90)90057-I.

Saaty, Thomas L. "Some mathematical concepts of the analytic hierarchy process." *Behaviormetrika* 18, no. 29 (January 1991): 1–9. https://doi.org/10.2333/bhmk.18.29_1.

Saaty, Thomas L. "Decision making with the analytic hierarchy process." *International Journal of Services Sciences* (2008): 83–98.

Sinha, G. K. "A project report on manufacture of automobile piston." *MSME Patna* (2012). https://msmedipatna.gov.in/Pprofile/Automobile%20Piston.pdf

Tahriri, Farzad, Rasid Osman, Aidy Ali, Rosnah Yusuff, and Alireza Esfandiary. "AHP approach for supplier evaluation and selection in a steel manufacturing company." *Journal of Industrial Engineering and Management* 1, no. 2 (2008): 54–76.

Tam, Maggie, and Rao Tummala. "An application of the AHP in vendor selection of a telecommunications system." *Omega* 29, no. 2 (2001): 171–182.

Varagas, Luis. "An overview of analytic hierarchy process: Its applications." *European Journal of Operational Research* 48, no. 1 (1990): 2–8.

Yadav, D., D. Chhabra, R. Kumar Garg, Akash Ahlawat, and Ashish Phogat. "Optimization of FDM 3D printing process parameters for multi-material using artificial neural network." *Material Today: Proceedings* 11, no. 225 (2019).

Chapter 6

Multi-Wire-Arc Additive Manufacturing (M-WAAM) of Metallic Components

Characterization and Mechanical Properties

V. Kumar, B.K. Roy, and A. Mandal

6.1 INTRODUCTION

Additive manufacturing (AM), also termed as "3D printing", is a process to fabricate 3D components by gradually adding suitable materials layer upon layer supported from a computer-aided design (CAD) model contradictory to that of the subtractive manufacturing process (Zhai et al., 2014). The term "additive manufacturing" is given by ASTM International's Committee F42 and is considered as an industrial-production technology. The process uses powder, wire, or sheets as feedstock materials to fabricate dense metallic objects by melting and solidification of molten material with the help of a laser beam, electron beam, or electric arc as an energy source to deposit the material in layer-upon-layer pattern (Yilmaz et al., 2016). The distinctive feature of AM is the production of medium to high complex shape structures without the use of expensive tooling such as casting molds, dies, and punches, and thus reduces the conventional processing steps (DebRoy et al., 2018). AM processes are further classified into seven broad areas based on the deposition technique, material types, and the way material is fused and solidified (Nath et al., 2020). The complete working phases of the AM and its classification have been presented in Figure 6.1. The chapter focuses on the basic principle of single-wire- and multi-wire-based additive manufacturing (M-WAAM), and its versatile application to fabricate various alloys and metal matrix composite (MMC) along with the study of morphological behavior and mechanical properties of the fabricated parts.

6.2 WORKING PRINCIPLE OF THE WAAM SYSTEM

Directed energy deposition (DED) is one of the families of AM processes that use a thermal heat energy source to melt and deposit the feedstock materials in the form of metallic powder or wire. Based on the type of energy source used, its subcategories are plasma arc, electron beam, laser beam, and electric arc–based DED (Dass et al., 2019). However, based on the type of feedstock material, it is classified as powder- and wire-based DED. The major area of application of DED-AM includes repairing and remanufacturing of the damaged parts to increase their lifespan and hence reduce the environmental impact (Wahab et al., 2019). But several advancements have been made to increase its area of utility and the researchers have come up with the hybrid process in which metallic wire is directly fed into a molten pool generated from the electric arc energy and the process is known as electric

DOI: 10.1201/9781003270027-6

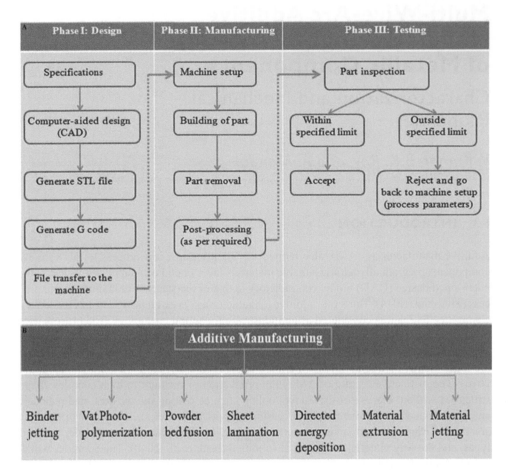

Figure 6.1 Additive manufacturing: different phases and its basic classification (Abdulhameed et al., 2019).

arc–based DED, popularly known as WAAM (Rodrigues et al., 2019). The arc welding is used as the energy source to fuse the metallic wire and fabricate metal parts additively in a layer-by-layer manner that is controlled by a robotic arm or a CNC system to direct the movement of the welding torch and wire feeder. The schematic diagram and experimental setup for a basic WAAM system have been shown in Figure 6.2. The key attraction of the WAAM for modern industries is the lower cost per component, shorter production lead time, and flexibility to build or repair complex structures (Dinovitzer et al., 2019). The higher deposition rate, good mechanical properties, and acceptable dimensional accuracy make it a suitable choice for building medium- to large-scale components over other similar AM processes (Khan et al., 2020). Depending on the component size to be produced, it reduces the deposition time by 40–60% and post-processing time by 15–20% compared to the traditional manufacturing process (Vaezi et al., 2020). In the recent advancement, the process development successfully leads to manufactured aircraft landing gear ribs by

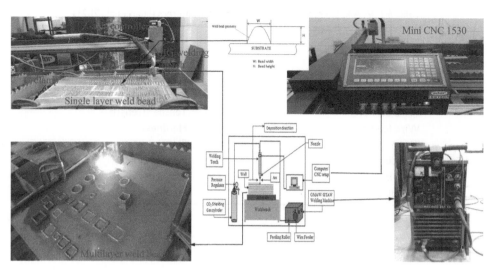

Figure 6.2 Schematic diagram and experimental setup for a basic WAAM system (Kumar et al., 2021).

cutting off the raw material closely to 78% when compared to the traditional machining process (Wu et al., 2018a, 2018b).

6.3   PROCESS VARIABLE AND CONTROL IN WAAM

The WAAM system is a complex manufacturing process regulated by various controllable variables. The use of a CNC machine or robotic arms for deposition and feeder nozzle movement makes it a complicated setup and thus required to control each variable parameter with great understanding to obtain optimal results while fabricating the components (Kozamernik et al., 2020). The prime concerns with this process are the morphology and deposit quality of the fabricated part that includes surface finish, distortion, and residual stress (Yehorov et al., 2019). A major part of this system is the welding machine and therefore aforementioned issues majorly rely on selected metallic wire properties and the processing input parameters (Teja et al., 2020; Liberini et al., 2017). The possible input and output process parameters are illustrated in Table 6.1. The inter-layer temperature during the WAAM process largely affects the cooling pattern of the deposited part, thus affecting the mechanical and microstructural characteristics of the fabricated part (Wu et al., 2018a). The high value improves the wettability of the molten pool leading to uniform deposition; however, a lower value results in component brittleness and poor bead-to-bead remelting responsible for porosity (Geng et al., 2017). Thus, it is important to maintain the inter-layer temperature at an appropriate range by providing inter-layer dwell time and preheating of the base plate. The heat transfer during the material deposition is mostly transient, non-equilibrium, and non-uniform, leading to induce residual stress and distortion in the built part (Ding et al., 2011). Therefore, thermal management during the process plays a crucial role for the quality product to be fabricated. The basic variant of this process includes GMAW-based, GTAW-based, and PAW-based WAAM systems. However, with the process advancement, various other variants come into

Table 6.1 WAAM Process Input and Output Parameters (Jafari et al., 2021)

Input Parameters	Output Parameters
Arc current	Surface finish
Welding voltage	Dimensional accuracy
Wire material	Residual stress
Wire diameters	Tensile strength
Wire feed rate	Hardness
Torch travel speed	Elongation
Shielding gas type	Bead geometry
Inter-layer dwell time	Distortion
Shielding gas flow rate	Inter-layer temperature

the picture that improves process performance during fabrication. They are cold metal trans-fer (CMT) based GMAW-WAAM, pulsed based PAW-WAAM, tandem GMAW-WAAM, and variable polarity (VP) based GTAW-WAAM. Diversity of the WAAM review works has been published by the researchers that cover state-of-the-art systems, design, and control of process parameters, material quality, and performance, area of application, sensing, and monitoring, path planning strategy, and so on (Dass et al., 2019; Dhinakaran et al., 2020; Rosli et al., 2021). Nevertheless, there is still a need for a systematic review of the work based on the mechanical and microstructural characterization of components fabricated by double- or multi-wire-based WAAM techniques.

6.4   MULTI-WIRE PLUS ARC ADDITIVE MANUFACTURING (M-WAAM): PROCESS MODIFICATION

The requirement to manufacture different alloys, metal matrix composite (MMC), and func-tionally graded composite materials (FGMs) leads to a new advancement in the area of WAAM that uses more than one metallic wire, feeding simultaneously to a common molten pool to fabricate the desired part (Reisgen et al., 2019). This modification in the WAAM setup widens the scope and its utility in the various industries. The used technique to fabri-cate desired alloy or metallic composite is suitable with welding techniques such as GTAW, CMT-based GMAW, or PAW-based WAAM. The use of multi-wires in the WAAM system increases the deposition rate to a great extent, thus making the process faster even for fabri-cating larger size components (Martina et al., 2019). The schematic diagram to differentiate single- and multi-wire-based WAAM is shown in Figure 6.3. With this development, many of the researchers worked on layering the component using different wire materials to add on desired characteristics features to the substrate required to increase the lifespan even in a critical environment. Besides this, work has been progressed to fabricate components with alloying materials or FGMs and study their mechanical and microstructure behavior to understand the feasibility of multi-wires in the WAAM system. The major advantage of the multi-wire arc additive manufacturing over the traditional single-wire-based WAAM is

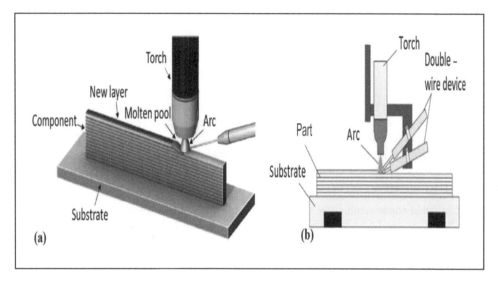

Figure 6.3 Schematic diagram of GTAW-based (a) single-wire and (b) double-wire feed WAAM system.

Source: Reprinted from Rodrigues, Tiago A. et al. *Materials* 12.7 (2019).

the high deposition efficiency, reduced heat input, porosity, and number of inclusions (Ren et al., 2021).

6.5 FABRICATION OF ALLOY THROUGH MULTI-WIRE-BASED WIRE ARC AM PROCESS (M-WAAM): ALUMINUM-BASED ALLOY

Aluminum is one of the most abundant metallic elements comprising 8% of the earth's crust and the third most common element after oxygen and silicon on our planet (Sharma et al., 2015). Lightweight, high strength-to-weight ratio, and corrosion-resistive properties are some of the versatile properties of Al for which it is widely used in many industries after steel. However, the welding of materials like aluminum and its alloys has always been troublesome, because of its reactivity toward atmospheric oxygen (Arun et al., 2015).

6.5.1 Al-Cu-Mg (Al 2024 Alloy)

Al 2024 is commonly used in industries like aerospace and defense. However, such variants of Al wire with a similar composition to Al 2024 are hard to find and so difficult to manufacture directly using the traditional WAAM process with a single metallic wire. Therefore, Z. Qi et al. have fabricated thin-wall components of 2024 Al alloy using two dissimilar 1.2-mm filler wires of ER2319 (an alloy of Al and Cu) and ER5087 (an alloy of Al and Mg) deposited simultaneously over the Al substrate plate (Qi et al., 2019). To obtain identical elemental composition to that of Al2024 alloy, the wire feed rate (WFR) of both the wires needs to be calibrated using Equation (6.1) (Qi et al., 2019):

$$E = \frac{\sum WFR_i D_i^2 \rho_i E_x}{\sum WFR_i D_i^2 \rho_i} \qquad (6.1)$$

where E_x is the element (x) content in the metallic wire, WFR_i $(i = 1, 2)$ is the wire feed rate, and ρ_i and D_i are the material density and wire diameter, respectively. The optical emission spectrometer (OES) has been used to identify the composition of the fabricated sample and found very similar to that of calculated compositional results.

6.5.1.1   Microstructures

The fabricated components were treated under three conditions: as-deposited (no heat treatment [HT]), proceed with HT (solution treatment with natural aging) T4, and HT (solution treatment with artificial aging) T6. These samples with (T4, T6) and without HT were further tested for the study of morphology and mechanical properties. The result indicates that mostly dendrites structure with fine equiaxed and columnar grain developed over the fabricated part that gets vanished after the HT and the existing phase gets transformed to α-Al+Al₂Cu from α-Al+Al₂Cu+Al₂CuMg confirmed from the XRD phase analysis, as shown in Figure 6.4.

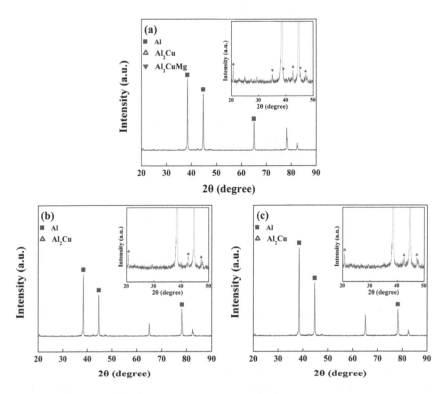

Figure 6.4 XRD plot for WAAM 2024 alloys: (a) as-deposited; (b) T4 heat-treated; (c) T6 heat-treated.

Source: Reprinted from Qi, Zewu et al. *Journal of Manufacturing Processes* 40 (2019): 27–36.

6.5.1.2   Mechanical Properties

The averaged micro-hardness of the fabricated 2024 Al alloy was examined and found to be 95 HV that gets enhanced with post-deposited HT (T4, T6). The tensile strength of the fabricated alloy was found to be isotropic in nature and gets anisotropic after HT. The results from the mechanical testing of the fabricated sample have been shown in Figure 6.5. Along the horizontal direction, the strength and elongation of the WAAM-deposited alloy has been improved because of HT; however, elongation along the vertical direction of the deposition gets decreased.

6.5.2   Al-Mg-Si Aluminum Alloy

Thin-wall components of ternary Al alloy (Al-3.1Mg-2.0Si) have been fabricated by Qi et al. (2018) through a double-wire-based WAAM process using the concept of synchronous feeding of ER5087 and ER4043 wires. The VP-GTAW-based power source with a current value of +100 A/–120 A and DCEP:DCEN = 1:4 were used for melting the metallic wire. The WFR for ER5087 was taken as 2.4 m/min and for ER4043 as 1.5 m/min. The torch travel speed of 300 mm/min, an arc length of 5 mm, and a constant rate of shielding gas (99.99% Ar) of 15 l/min were used during experimentation.

6.5.2.1   Microstructure

The morphological study on WAAM-fabricated Al-3.1Mg-2.0Si alloy showed the presence of α-Al, Mg_2Si, and Al_9Si as the main existing phase. The phase α-Al was found mostly in the alloy matrix, while the phases Mg_2Si and Al_9Si were uniformly distributed around the grain boundary. The SEM image of the fabricated alloy sample has been shown in Figure 6.6.

6.5.2.2   Mechanical Properties

The fabricated Al-3.1Mg-2.0Si alloy was tested for microhardness and found to be 54.7 HV. The plot for micro-hardness and tensile properties has been shown in Figure 6.7. The yield strength (YS), elongation, and ultimate tensile strength (UTS) of the deposited alloy along the horizontal direction were found to be 76.6 MPa, 11.4%, and 176 MPa, respectively, and observed to be marginally lesser when measured along the vertical direction.

6.5.3   Al-Mg-Sc Aluminum Alloy

The authors made a comparative study on the fabricated alloy with Al-6.0Mg-0.3Sc metallic wire employing a twin-wire and single-wire plus arc AM process (Ren et al., 2020). The experimental setup for the double-wire feed WAAM system and the fabricated sample with a single- and double-wire feed WAAM system has been shown in Figure 6.8. The metallographic structure, phase analysis, and mechanical properties of the fabricated samples were studied and a comparison has been drawn between the two processes. The cold metal transfer + pulse (CMT + P) process has been used as a heat source and Ar gas with a flow rate of 25 l/min for the shielding purpose. According to the heat input as shown in Equation (6.2), the instantaneous heat input from single-wire and each wire of twin-wire arc forming was

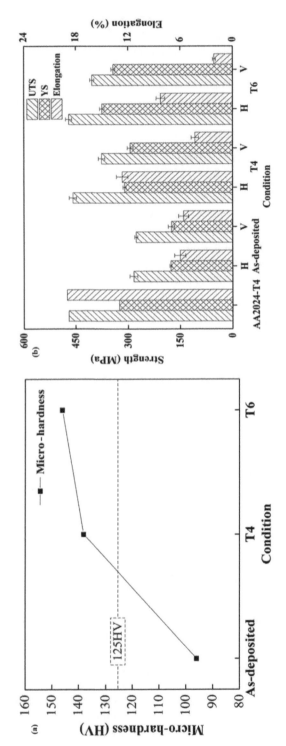

Figure 6.5 Averaged micro-hardness (a) and tensile properties (b) of WAAM-fabricated 2024 alloys under different conditions.

Source: Reprinted from Qi, Zewu et al. *Journal of Manufacturing Processes* 40 (2019): 27–36.

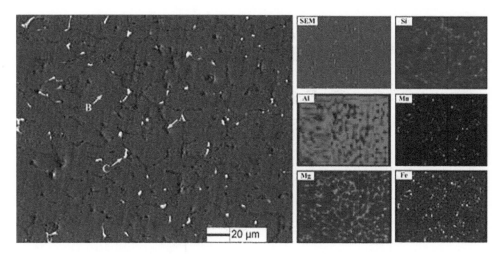

Figure 6.6 Microstructure and elemental distribution in the WAAM-fabricated Al-3.1Mg-2.0Si aluminum alloy.

Source: Reprinted from Qi, Zewu et al. *Materials Letters* 233 (2018): 348–350.

obtained as 149.328 J/mm, and 56.84 J/mm, respectively. Thus, arc of the twin-wire WAAM process has a lower heat input for the fabrication of the same component (Ren et al., 2020).

$$\text{Heat input}\left(\text{HI}\right) = \frac{\eta VI}{\vartheta_{TS}} \tag{6.2}$$

where I and V are the mean welding current and voltage, respectively; TS is the welding speed; η is the energy utilization factor taken as 0.8 for the CMT welding.

6.5.3.1 Microstructure

The microstructure and EDS result of the WAAM-fabricated sample using single- and double-wire feed has been shown in Figure 6.9. Relative to single-wire, the double-wire arc-formed deposits surface was observed to be more uniform and precise. The double-wire feed deposit exhibits lesser and smaller size pores with lower heat yields. The amount of precipitate $\beta(Mg_2Al_3)$ and $(FeMn)Al_6$ phases were fewer and smaller in size inferred from Figure 6.9a and b. The solid solution of Mg and Mn increases in the Al matrix of double-wire-based WAAM deposits and enhances the alloy strength, and a smaller primary Al_3Sc phase has been observed in Figure 6.9c responsible for excellent grain refinement. Furthermore, the presence of higher amount of Sc dissolved in the Al matrix due to lower heat input to the process has been confirmed from the EDS result shown in Figure 6.9d. The melting of rear wire "heat-treats" the matrix formed by the front provides energy for the precipitation of secondary Al3Sc phase that further strengthens the deposited alloy using double-wire feed WAAM process.

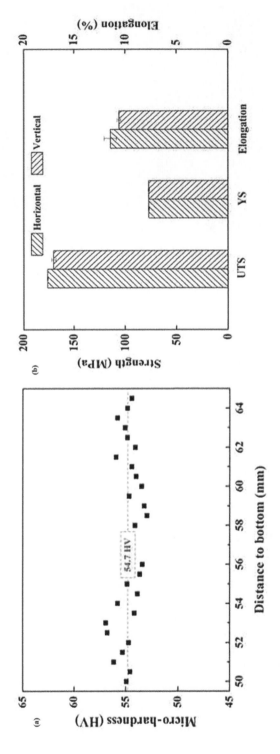

Figure 6.7 Micro-hardness plot (a) and tensile properties (b) of the fabricated Al-3.1Mg-2.0Si alloy.

Source: Reprinted from Qi, Zewu et al. *Materials Letters* 233 (2018): 348–350.

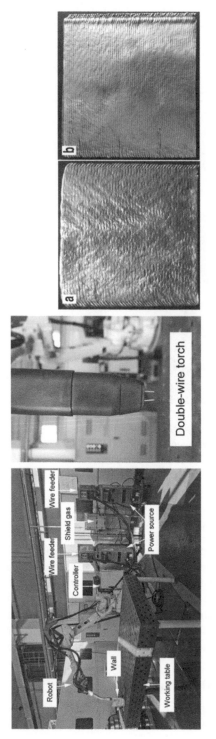

Figure 6.8 Double-wire-based WAAM system and component built with (a) single- and (b) double-wire feed.

Source: Reprinted from Ren, Lingling et al. *3D Printing and Additive Manufacturing* (2021).

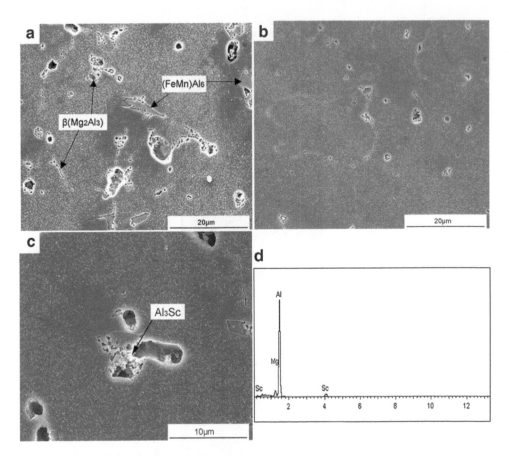

Figure 6.9 SEM image: (a) single-wire feed deposits, (b) double-wire feed deposits, (c) and (d) morphology and EDS results of primary Al3Sc phase.

Source: Reprinted from Ren, Lingling et al. *3D Printing and Additive Manufacturing* (2021).

6.5.3.2   *Mechanical Properties*

The yield strength, tensile strength, and % elongation were estimated as 258MPA, 363 MPa, and 26%, respectively, for the double-wire arc-formed alloy, as shown in Figure 6.10. The mechanical properties of the horizontal and vertical specimens were observed to be consistent because of lesser number of minute size pores and their discrete distribution throughout the deposited part. The overall result shows an improvement in the mechanical properties of the double-wire arc-formed alloy because of smaller grain size and increased amount of Mn and Mg dissolved in the matrix along with the presence of secondary Al_3Sc phase precipitate in the Al matrix.

6.5.4   **Al-Zn-Mg-Cu Aluminum Alloy**

Three different metallic wires ER5356, ER2319, and Zn (99.99 %) were used by researchers to deposit and fabricate AA7050 alloy and study the effect of multi-wires melting and

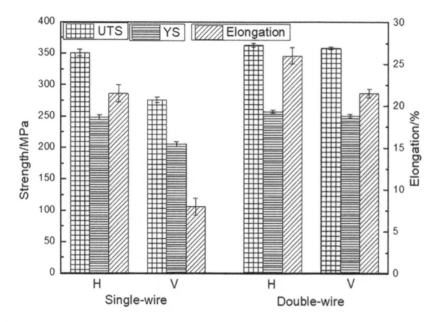

Figure 6.10 Mechanical properties of the fabricated alloy using single- and double-wire feed.

Source: Reprinted from Ren, Lingling et al. *3D Printing and Additive Manufacturing* (2021).

solidified at a time (Yu et al., 2021). The modified version of the GTAW-based WAAM system for multi-wire deposition has been used. The experimental designs with the wire feeding angle and fabricated sample have been shown in Figure 6.11. The stand-off distance for deposition was taken at 5 mm and the flow rate of Ar gas used for the shielding purpose was maintained at 15 l/min. The Zn wire kept in the middle of the other two metallic wires and deposits simultaneously into the weld pool with a constant feed rate and inter-layer dwell time of 180 seconds. The chemical composition of metallic wire and its consumed volume during the process are evaluated by Equations (6.3) and (6.4) (Yu et al., 2021):

$$E\% = \frac{\sum V_i \rho_i E_i}{\sum V_i \rho_i} \tag{6.3}$$

$$V_i = \pi r_i^2 \times WFS_i \tag{6.4}$$

where $E\%$ is the elemental composition of the fabricated alloy in wt.%; ρ_i (g/cm³) and V_i (cm³) are respectively the wire density and volume of metallic wire consumed while melting. The E_i (wt.%) represents the mass fraction of the element in the metallic wire and r_i is the wire radius (mm).

6.5.4.1 Microstructure

A morphological study on the fabricated sample stated that phase composition mostly comprises α-Al, η(MgZn₂), S(Al₂CuMg), and θ(CuAl₂) phases. The elemental composition of

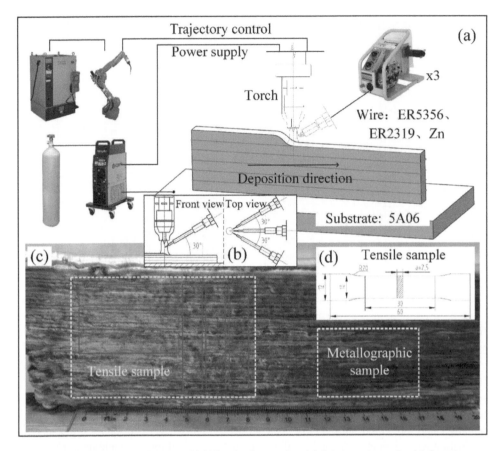

Figure 6.11 (a) Experimental design. (b) Wire feeding angles. (c) Fabricated sample. (d) Sample standard for tensile testing.

Source: Reprinted from Yu, Zhanliang et al. *Journal of Manufacturing Processes* 62 (2021): 430–439.

the fabricated parts was found to be in accordance with the compositional range of the Al 7050 alloys. The SEM image and elemental mapping of the scanned area are presented in Figure 6.12. The upper and bottom regions of the fabricated part were observed to have equiaxed and columnar crystals structure, respectively. Few defects like smaller size pores and micro-cracks were also observed in the inter-layer of the deposited parts.

6.5.4.2 Mechanical Properties

The micro-hardness of the fabricated alloy samples was found to be in the range of 95–115 HV. The mean hardness of the sample was observed to increase with an increase in travel speed; however, it gets decreased with an increase in arc current from 110 to 130 A. The tensile strength measured about the horizontal and vertical directions of deposition was found to be 241 and 160 MPa, respectively. The plot describing the range of micro-hardness and tensile properties has been shown in Figure 6.13.

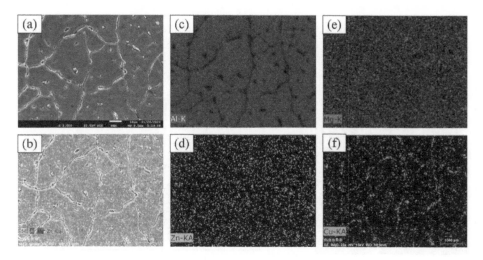

Figure 6.12 (a) SEM scan image of the fabricated sample. (b) General view of map scanning. (c–f)
Elemental distribution of Al, Zn, Mg, and Cu in the sample.

Source: Reprinted from Yu, Zhanliang et al. *Journal of Manufacturing Processes* 62 (2021): 430–439.

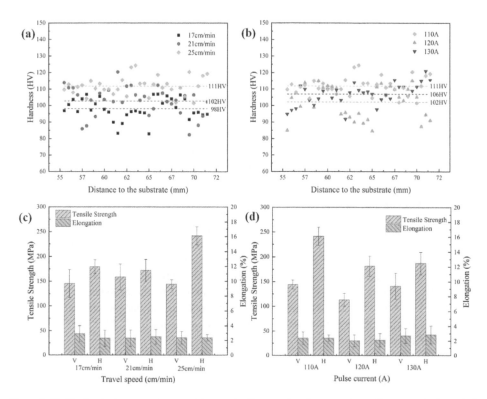

Figure 6.13 Mechanical properties: travel speed effect on hardness and tensile strength (a and c),
pulse current effect on hardness and tensile strength (b and d).

Source: Reprinted from Yu, Zhanliang et al. *Journal of Manufacturing Processes* 62 (2021): 430–439.

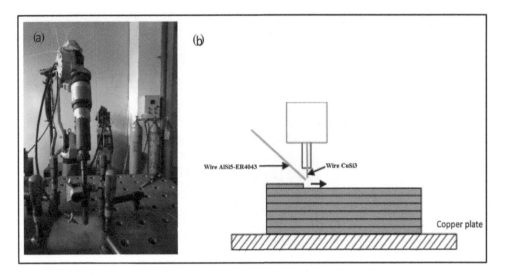

Figure 6.14 (a) CMT-based WAAM setup. (b) Schematic diagram for wire feeding.

Source: Reprinted from Wang, Yanhu et al. *Applied Surface Science* 487 (2019): 1366–1375.

6.5.5 In Situ Cu-Al-Si Alloy

The CMT-based WAAM system with an additional wire feeder was used to produce Cu-Al-Si alloy (Wang et al., 2019). An advanced 4000R NC controller CMT-based GMAW was used as an energy source to melt and feed 1.2-mm $CuSi_3$ and $AlSi_5$ wires to fabricate the alloy. The basic experimental setup has been shown in Figure 6.14. The complete work has been divided into three parts: fabricate the two samples with varying Al content in the alloy, study the influence of silicon on the properties of fabricated Cu-Al alloys, and study performance of the fabricated alloys using the morphological and mechanical examination.

6.5.5.1 Microstructures

The microstructure investigation shows that abundant columnar grains developed along and perpendicular to the boundary layers with few of the large epitaxial columnar grains developed toward the stacking direction. For both the samples (1, 2), ingredients gradually become homogenized along the deposition direction. The optical image of both samples has been shown in Figure 6.15. The XRD phase analysis confirms the presence of four phases in sample 1 as SiO_2, Cu, Cu_9Al_4, and $CuAl_2$. Sample 2 has similar phases that exist with the minor variation of the planes and intensity for the constituent $CuAl_2$ depending on the amount of Al and Cu.

6.5.5.2 Mechanical Properties

Both samples 1 and 2 built by varying Al percentages resulted in better strength and ductility. The hardness value was found to increase with an increase in Al content in the alloy. Also, the result

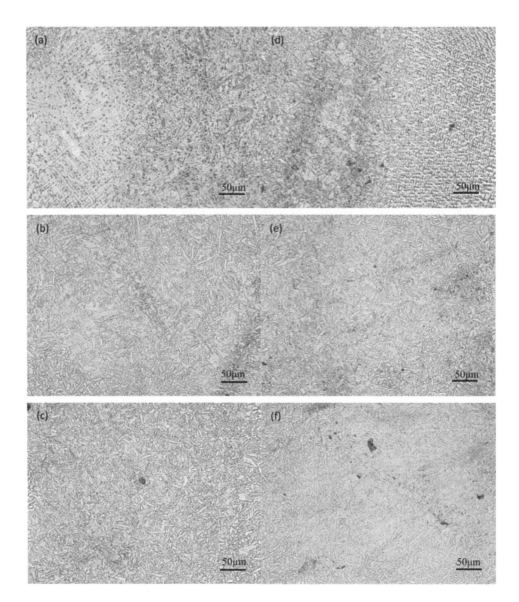

Figure 6.15 Optical microstructures of sample 1 at lower region (a), middle region (b), and upper region (c); and sample 2 at lower region (d), middle region (e), and upper region (f).

Source: Reprinted from Wang, Yanhu et al. *Applied Surface Science* 487 (2019): 1366–1375.

shows that the addition of a small quantity of silicon (2.1–2.4%) in the metallic wires effectively improves the hardness and tensile strength when compared to pure Cu-Al alloy. Sample 2 was observed to have a higher value of UTS and YS because of an increase in Al and Si content; however, a higher value of the elongation was observed in sample 1 compared to sample 2. The results for the mechanical properties of samples 1 and 2 have been plotted in Figure 6.16.

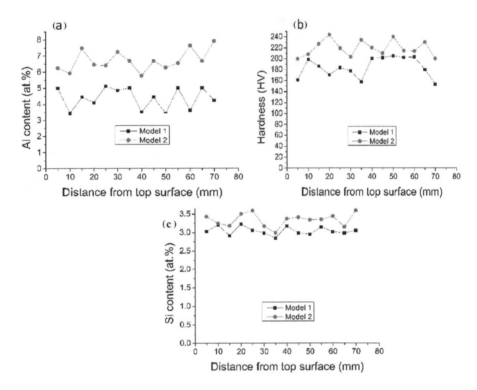

Figure 6.16 (a) Al content near the hardness testing points. (b) Hardness variation along the center-line. (c) Si content near the hardness testing points.

Source: Reprinted from Wang, Yanhu et al. *Applied Surface Science* 487 (2019): 1366–1375.

6.6 TITANIUM-BASED ALLOY

Titanium and its various alloys are increasingly being studied and workout in the WAAM, due to the reduction of higher costs associated with the fabrication of Ti and its alloy-based components with other traditional methods. The properties of these materials such as higher strength, greater toughness, better corrosion resistance, and ability to sustain at higher temperatures with minimal loss of its mechanical properties make it one of the favorable materials for aerospace industries and biomedical applications (Gloria et al., 2019).

6.6.1 Ti-6Al and Ti-6Al-7Nb Titanium Alloy

The basic principle to fabricate multi-element alloy based on the twin wire + arc AM process (T-WAAM) combined with in situ alloying has been adopted by the researchers (Yang et al., 2020). GTAW technique has been used as a heating source to melt and deposit Ti-6Al alloy by regulating the feed rate of both the metallic wires and in continuation fabricated a noble bio-medical Ti-6Al-7Nb alloy by addition of element Nb in the form of foil. The welding wires of 1.2 m diameter were used as a feedstock along with the Nb foil of 50 μm thickness. The combined T-WAAM and in situ alloying setup have been illustrated in Figure 6.17.

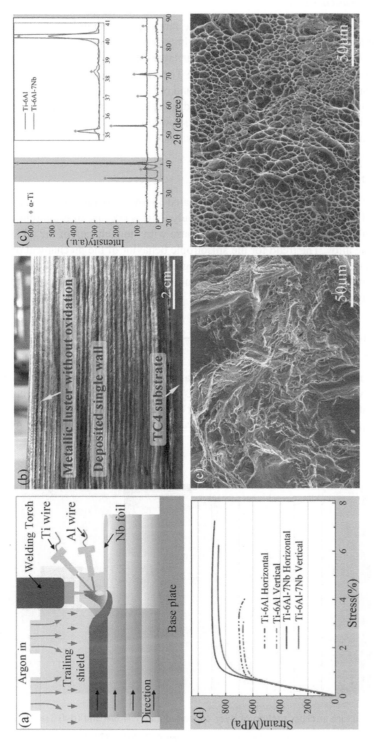

Figure 6.17 (a) Experimental setup for Twin-WAAM combined with in situ alloying. (b) Fabricated wall component of Ti-6AL-7Nb alloy. (c) XRD phase result. (d) Strain–stress curve for the built component in a different direction. (e) Fracture morphology of Ti-6Al. (f) Fracture morphology of Ti-6Al-7Nb.

Source: Reprinted from Yang, Zhenwen et al. *Materials Research Letters* 8.12 (2020): 477–482.

6.6.1.1   Microstructures

The morphological study has been done on both the fabricated titanium-based alloys. The Widmanstätten microstructure with basket weave impression has been observed for the deposited Ti-6Al-7Nb alloy. Except for the last four layers, the heat-affected zone (HAZ) with a bandwidth of 400–500 μm appeared for the Ti-6Al-7Nb alloy. The microstructure with Nb-rich β-phase in nanoscale appeared in the band of HAZ was finer than that appeared in the non-HAZ band. The microstructures developed at the cross section of deposited layers have been shown in Figure 6.18. The XRD results showed the presence of α and a lattice phase of closely packed hexagonal in both the deposited alloys.

6.6.1.2   Mechanical Properties

The mean value of UTS and elongation of the fabricated Ti-6Al-7Nb alloy were found to be higher than Ti-6Al alloy by 35% and 47%, respectively. This is because of grain refinement and due to the evolution of the nanoscaled lath β-phase with the inclusion of the additional Nb element into the alloy. The trend for the anisotropic behavior of the mechanical properties remains the same for both the alloys. The fractography studied on Ti-6Al alloy shows a typical brittle fracture property and on Ti-6Al-7Nb alloy shows a ductile fracture with zero appearance of cleavage.

6.6.2   Functionally Graded Titanium–Aluminide (TiAl) Alloys

J. Wang et al. fabricated a functionally graded material (FGMs) as TiAl alloys using a T-WAAM system with GTAW welding equipment as the power source and investigated the variation in microstructures and mechanical behavior by altering the alloy composition along the direction of deposition (Wang et al., 2018). The Pure Ti wire of 1.0 mm and 1080 Al wire

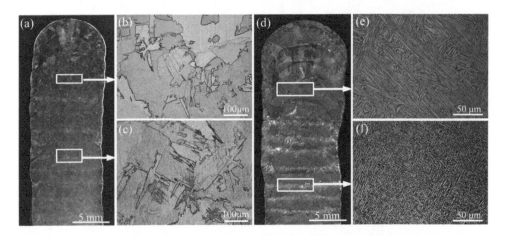

Figure 6.18 Microstructure at cross section of layer deposited: (a–c) Ti-6Al and (d–f) Ti-6Al-7Nb.

Source: Reprinted from Yang, Zhenwen et al. *Materials Research Letters* 8.12 (2020): 477–482.

of 0.9 mm diameter were melted into a common molten pool. The constituent ratios of both the wire were controlled by their feed rate. To keep the molten pool stable during wire feeding and deposition, the angle between the nozzle and the substrate plate was taken as 30°.

6.6.2.1   Microstructure

The morphological study was made on a different section of the deposited part and the image is shown in Figure 6.19. The result shows that with an increase in the Al content from the base to the upper section of the fabricated part, a layered structure in sequence consists of α–β duplex structure, α–α_2 lamellar structure, large α_2 grains, α_2–γ duplex lamellar structure, and γ inter-dendritic structure were observed. The volume fraction and microstructure of different phases that exist depend mostly on Al concentration in the deposited wall.

6.6.2.2   Mechanical Properties

With an increase in Al content along the deposition direction, tensile strength and micro-hardness of the alloy exhibit identical patterns, that is, at first they increase to a peak value and subsequently decrease. The results from micro-hardness testing on the fabricated sample have been shown in Figure 6.20. The peak value of both the parameters observed in the alloy with 30–33% Al content exists in a single α_2 phase. The oxidation resistance properties were observed to be degraded with decreasing Al content in the fabricated wall.

6.6.3   Titanium Alloy–Alloy Composites

The modified WAAM system has been developed to fabricate effective "Alloy–Alloy" composite (AAC) material that allows alternating wire feeding of pure Ti (CPTi) and Ti-64

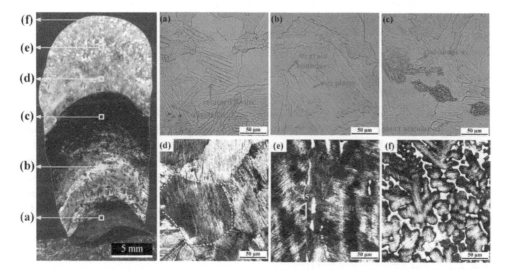

Figure 6.19 Microstructural evolution with increasing Al content from the bottom to the top.

Source: Reprinted from Wang, Jun et al. *Materials Science and Engineering: A* 734 (2018): 110–119.

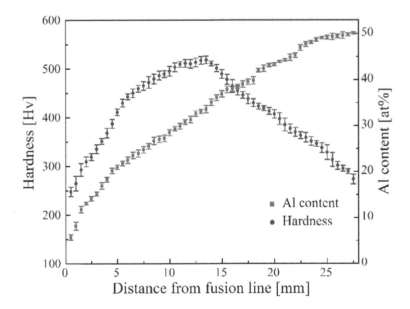

Figure 6.20 Al content and micro-hardness testing along the centerline from the fusion line to the top surface.

Source: Reprinted from Wang, Jun et al. *Materials Science and Engineering: A* 734 (2018): 110–119.

metallic wire of 1.2 mm diameter (Davis et al., 2019). The AAC WAAM wall of size 350 × 100 × 32 mm was built with 60 layers of deposition following the 50:50 volume ratio of both the welding wire deposited. Plasma arc welding setup has been used as an energy source to melt the wire. The investigation suggests it as a novel method for printing AAC material of the varying composition. The schematic diagram for the bead profile and the deposited AAC wall has been shown in Figure 6.21.

6.6.3.1   Microstructure

A morphological study on the WAAM-deposited AAC wall observed to have refined and diverse grain structure consists of very small columnar β-grains in comparison to that found in a traditionally built wall of Ti64 with few isolated and equiaxed grains. The position of such isolated and equiaxed grains is linked with the transition in alloy composition between solute-rich and lean deposition tracks. The optical micrograph of the fabricated ACC wall component has been shown in Figure 6.22 for the WD-ND and TD-ND planes. A regular "fish scale"-like pattern was observed in the macrograph for the distribution of Al originated by overlapping pure Ti and Ti64 alternate alloy tracks.

6.6.3.2   Mechanical Properties

The results from the mechanical investigation showed that the TS, YS, and UTS of the WAAM-fabricated ACC wall were found to have a lower value than that of the conventionally

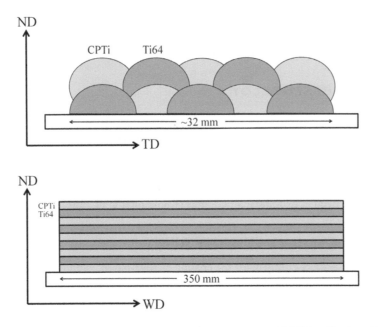

Figure 6.21 Schematic diagram for WAAM deposition strategy to build ACC wall, where WD = heat source moves parallel to the wall, TD = wall transverse direction, and ND = build height direction perpendicular to each layer.

Source: Reprinted from Davis, A. E. et al. *Materials Science and Engineering: A* 765 (2019): 138289.

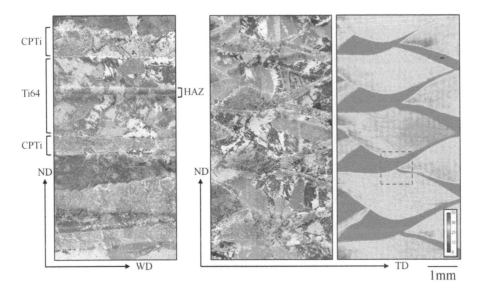

Figure 6.22 Optical micrograph cross sections of the as-deposited AAC-WAAM component are presented.

Source: Reprinted from Davis, A.E. et al. *Materials Science and Engineering: A* 765 (2019): 138289.

fabricated Ti64 wall component. In the loading direction, AAC fabricated material has an edge of higher yield stress parallel to the track alignment but has a limitation of advancing significant plastic heterogeneity in the transverse loading direction. The stress–strain curve for the fabricated sample has been shown in Figure 6.23.

6.7 NICKEL-BASED ALLOY

Nickel, a non-ferrous metal, is found in abundance in the earth's crust. Nickel with its different alloys possesses excellent properties such as higher strength and toughness value, exceptional corrosion resistance property, and the ability to retain properties even at elevated temperatures (Ezugwu et al., 1999). Inconel 625 and Inconel 718 are some of the popularly known Ni-based superalloys used in the aviation industry because of their property to retain higher strengths even at extreme temperatures (Ramkumar et al., 2017).

6.7.1 NiTi Coating on Ti6Al4V Alloys

J. Wang et al. have successfully made the coating of NiTi-based composite over the titanium-based Ti64 substrate through varying arc currents in the range of 50–70 A using GTAW-based WAAM system composed of a dual-wire feeder that allows the in situ alloying from two separate wire of pure Ni of 0.9 mm and pure Ti of 1.2 mm diameter (Wang et al., 2020). The schematic representation of the double-wire feed WAAM system has been illustrated in Figure 6.24. The angle between the feeder nozzles to that of the base plate was taken as 30° to assure the stability of the molten pool during wire melting and deposition. The ratio of Ti:Ni in the alloy coated over the substrate plate maintained as 55:45 was done by adjusting the WFS of both the metallic wire.

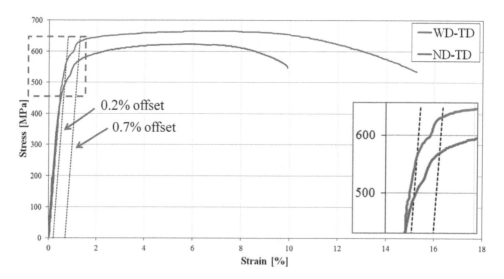

Figure 6.23 Stress–strain curves for the AAC WAAM sample cut with the gauge length along the heat-source travel direction and gauge length aligned with the vertical build-height direction.

Source: Reprinted from Davis, A. E. et al. *Materials Science and Engineering: A* 765 (2019): 138289.

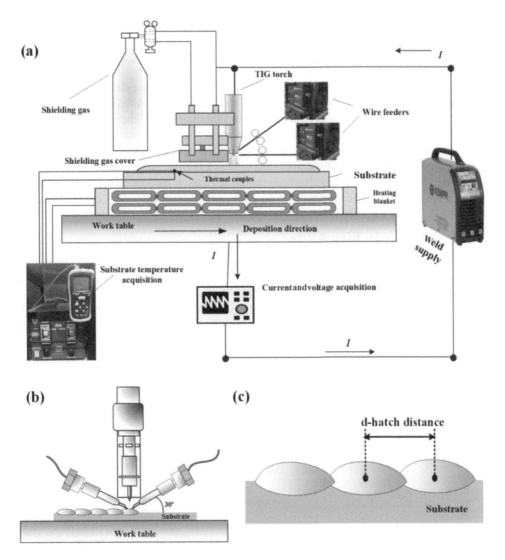

Figure 6.24 Schematic diagram for (a) double-wire-based WAAM process, (b) nozzles for double-wire feed and its deposition angle, and (c) coating sketch map.

Source: Reprinted from Wang, Jun et al. *Surface and Coatings Technology* 386 (2020): 125439.

6.7.1.1 Microstructure

The morphological investigation was made on the sample coated with NiTi-based composite using the dual-wire-based WAAM technique. The results come up with the phase $NiTi_2$ distributed throughout the tested sample together with few NiTi dendrites coarse structures or fine α-Ti dendrites subjected to welding current provided during deposition. The coating of the alloy at 50 A comprises NiTi coarse dendrites structure as the minor phase, while coatings at an arc current of 60 and 70 A comprise α-Ti fine dendrites structure as the minor phase. However, the major part still comprises NiTi$_2$ phase in all the arc current settings. The backscattered electron imaging of the fabricated sample under different current values has been shown in Figure 6.25.

6.7.1.2 Mechanical Properties

Three samples with NiTi coating at three different arc current setting during deposition have been investigated. The result shows that with an increase in current values from 50 to 70 A, the mean value of coating thickness got improved from value 1.56 to 1.91 mm due to an increase in dilution rate between the base plate and coating material. The micro-hardness values of the NiTi coatings over the substrate plate were found to be higher than the uncoated Ti-64 base plate. The mean value of micro-hardness for corresponding arc current 50, 60, and 70 A resulted as 715, 818, and 758 HV0.2, respectively, has been shown in Figure 6.26. Among the three arc current settings, the best results were obtained for the NiTi coating deposited at 60 A current values.

6.7.2 Ni$_{53}$Ti$_{47}$ In Situ Alloy

In order to minimize the fabrication cost of NiTi alloy using a conventional technique, an innovative double-wire-arc AM process along with in situ alloying has been applied by C. Shen et al. to fabricate polycrystalline Ni$_{53}$Ti$_{47}$ (Shen et al., 2020). The two metallic wires of pure Ni and Ti (0.9 mm diameter) have been used to design and fabricate the desired alloy during the in situ WAAM process. For deposition, the arc current and wire travel speed were taken as 140 A and 95 mm/min and thus maintained the specific deposition energy during the in situ alloying to 19.52 kJ/g. A total of 35 layers of deposition have been done to build up the sample wall. The configuration of the dual-wire deposition torch used during fabrication has been shown in Figure 6.27.

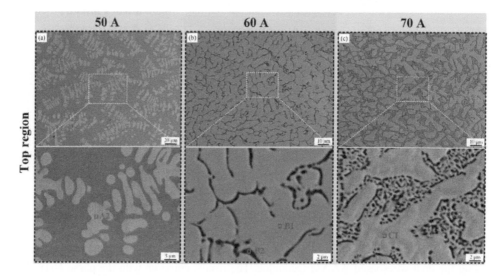

Figure 6.25 BSE images taken at different regions of the coatings deposited under different arc current intensity. (a) Top region under 50 A. (b) Top region under 60 A. (c) Top region under 70 A.

Source: Reprinted from Wang, Jun et al. *Surface and Coatings Technology* 386 (2020): 125439.

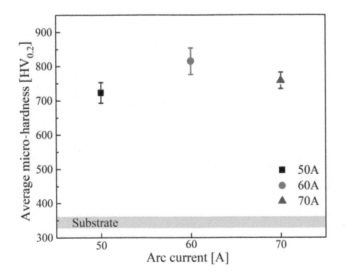

Figure 6.26 Mean hardness values of the coatings at different arc currents.

Source: Reprinted from Wang, Jun et al. Surface and Coatings Technology 386 (2020): 125439.

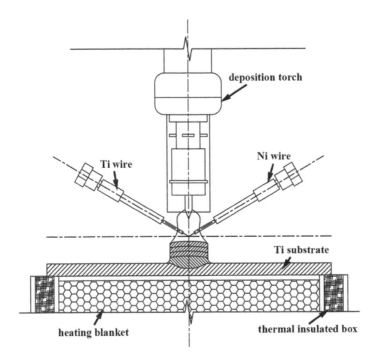

Figure 6.27 Schematic representation of the double-wire-based WAAM system for in situ alloying of NiTi alloy.

Source: Reprinted from Shen, Chen et al. Journal of Alloys and Compounds 843 (2020): 156020.

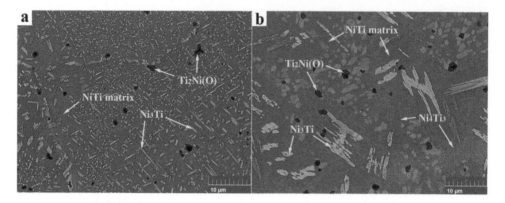

Figure 6.28 SEM micrographs of fabricated Ni53Ti47 alloy. (a) AF sample. (b) HT sample.

Source: Reprinted from Shen, Chen et al. *Journal of Alloys and Compounds* 843 (2020): 156020.

6.7.2.1   *Microstructure*

Neutron diffraction test has been conducted on the heated sample to characterize the microstructure and different phases obtained in the fabricated $Ni_{53}Ti_{47}$ alloy and provide information regarding the lattice transformation. It was found that the fabricated alloy was composed of NiTi and Ni_3Ti structures due to rapid solidification and short WAAM buildup time. However, when the deposited alloy is subjected to post–heat treatment (annealing), it confirms the presence of the metastable Ni_4Ti_3 phase generated and is responsible to increase the micro-strain in the NiTi phase, thus helping in martensitic transformation and elasticity. The SEM microstructure of the fabricated sample before and after heat treatment can be seen in Figure 6.28.

6.7.2.2   *Mechanical Properties*

The residual stress induced in the deposited NiTi alloy was found to be tensile in nature and were released during post–heat treatment at 528 °C. The rate of thermal expansion for NiTi and Ni_3Ti decreases during stress relief treatment. The neutron Rietveld refinement method was used to quantify the coefficient of thermal expansion for the NiTi and Ni_3Ti as 0.040×10^{-3} °C^{-1} and 0.036×10^{-3} °C^{-1}, respectively.

6.8   STEEL

Steel, an alloy of iron and carbon, with other additives possesses great importance as the manufacturing and construction material. It contributes to every aspect of life whether it is construction products, automobile sector, medical equipment, electronic products like refrigerators or washing machines, cargo ships, and many more (Vaghani et al., 2014). For more than 70 years, steel has been used with great success in various industries due to its versatile properties such as high strength, easy to fabricate, low maintenance, durable, long-lasting, appealing appearance, eco-friendly, and recyclable material (Bajaj et al., 2020).

6.8.1   Cr-Ni Stainless Steel

Researchers designed an innovative and highly efficient method based on a twin-wire feed plasma arc plus AM system (DWF-PAM) to manufacture Cr-Ni stainless steel alloy components (Feng et al., 2018). The modified version comprises GTAW-based power source, a plasma torch with a water-cooled channel, and two automatic wire feeders operating at a constant feed rate. The experimental setup has been shown in Figure 6.29. The H00Cr21Ni10 metallic wire of 1 mm diameter as the feedstock material and a base plate of 08Cr19Ni9 stainless steel have been used. The Ar (99.99%) gas is used for shielding the molten pool and for plasma generation to fabricate components with 70 deposited layers. The result shows an increase of deposition rate by 1.06 times to that of a single wire feed AM system (SWF-PAM) keeping the same process parameters in both cases.

6.8.1.1   Microstructure

The microstructural study on the interface zone of the deposited layers using the DWF-PAM process confirms the presence of highly developed equiaxed ferrite, while underdeveloped equiaxed ferrite was discovered in the same region for the SWF-PAM fabricated sample. The average grain size of parts built by the DWF-PAM process at three different feed rates was measured as 11.67, 9.52, and 8.96 μm, respectively, found to be much smaller in comparison to the grain size of the sample fabricated using the SWF-PAM process. The results of the morphological study have been shown in Figure 6.30.

Figure 6.29 The plasma arc plus additive manufacturing (AM) system.

Source: Reprinted from Feng, Yuehai et al. *Journal of Materials Processing Technology* 259 (2018): 206–215.

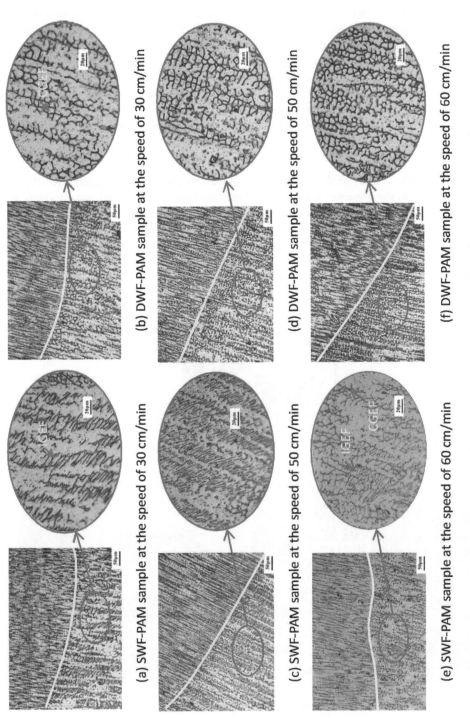

(a) SWF–PAM sample at the speed of 30 cm/min

(b) DWF–PAM sample at the speed of 30 cm/min

(c) SWF–PAM sample at the speed of 50 cm/min

(d) DWF–PAM sample at the speed of 50 cm/min

(e) SWF–PAM sample at the speed of 60 cm/min

(f) DWF–PAM sample at the speed of 60 cm/min

Figure 6.30 Microstructures of the fabricated sample by single- and double-wire feed PAM at different deposition speeds.

Source: Reprinted from Feng, Yuehai et al. *Journal of Materials Processing Technology* 259 (2018): 206–215.

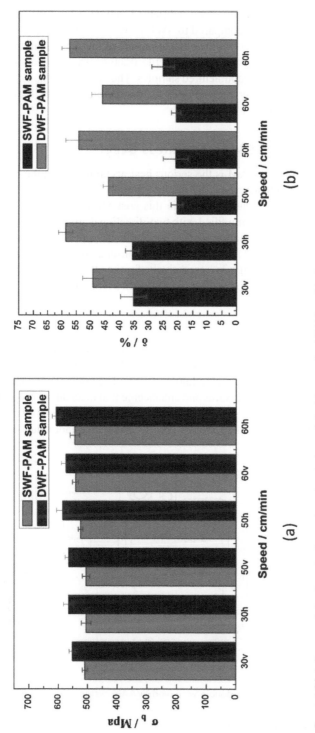

Figure 6.31 Influence of the deposited speed on the single- and double-wire feed PAM on (a) ultimate tensile strength and (b) percentage elongation.

Source: Reprinted from Feng, Yuehai et al. *Journal of Materials Processing Technology* 259 (2018): 206–215.

6.8.1.2 *Mechanical Properties*

The study on mechanical properties of the fabricated sample using DWF-PAW shows an increase in UTS by 10.2% when compared to SWF-PAW fabricated sample. Also, an increment of 176% has been confirmed for the elongation rate of the sample built by DWF-PAM to that of the SWF-PAM process. A typical ductile fracture was also confirmed from the fractographic study in both the manufactured samples. The bar graphs showing the variation in mechanical properties at different feed rates for the double- and single-wire feed PAM process have been shown in Figure 6.31.

6.8.2 Double-Wire SS316L Stainless Steel

The major limitation of the WAAM is the serious heat accumulation in the fabricated parts that restrain the process efficiency and its implementation in the metal additive manufacturing sector. Therefore, researchers worked in this area and developed a CMT + P mode dual-wire-based WAAM system modified with two-directional auxiliary gas attachment and studied its performance by fabricating parts with double-wire of SS316L stainless steel of 1.2 mm diameter (Wu et al., 2021). The values for arc current, voltage, and travel speed were taken as 103 A, 16.8 V, and 1.2 m/min, respectively, for the alloy deposition. The schematic diagram of the experimental setup has been presented in Figure 6.32.

6.8.2.1 *Microstructure*

The optimal value for different specifications of the auxiliary gas was obtained as 17.4° (nozzle angle), 25 l/min (gas flow rate), and 10.44 mm (SOD) with the use of Box-Behnken design model and response surface methodology (RSM). The optimal parameters resulted in morphology with refined grain structures and enhancement in mechanical properties because of the stirring action of auxiliary gas and its cooling effect during the fabrication. The flow of auxiliary gas stirs the unsolidified molten pool to protect the fabricated layers from the

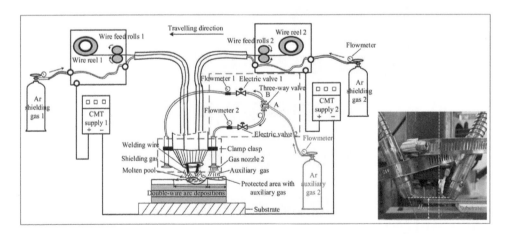

Figure 6.32 Schematic drawing of the double-wire feed WAAM system and the position of auxiliary gas nozzle and welding gun.

Source: Reprinted from Wu, Wei et al. *Metals* 11.2 (2021): 190.

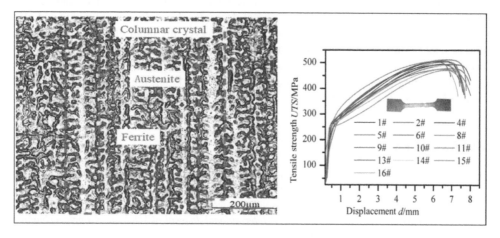

Figure 6.33 Microstructure of fabricated samples and tensile results of specimens [45].

Source: Reprinted from Wu, Wei et al. *Metals* 11.2 (2021): 190.

surrounding atmosphere and prevent their oxidation. The microstructure of the deposited layers confirmed the presence of columnar grain crystals composed of austenite and ferrite, as shown in Figure 6.33.

6.8.2.2   Mechanical Properties

The nozzle angle of 30° for the auxiliary gas torch results in poor morphology but developed exceptional properties for strong stirring action and advancing the molten pool forward. Also, increasing the auxiliary gas flow rate to 25 l/min increases the flow strength and stiffness that ultimately boost the gas stirring action on the molten pool and resist the interference. In comparison to the sample fabricated without the auxiliary gas process, the hardness, tensile strength, and elongation of the fabricated sample in presence of auxiliary gas were improved.

6.9   CONCLUSION AND FUTURE PERSPECTIVES

WAAM is an interdisciplinary technology that includes different welding techniques, materials science, thermomechanical engineering, computer-aided design, and control system. Therefore, there is plenty of scope in different areas for improving and introducing new modifications. Wire-feed AM is a promising technology for producing larger size components with low to moderate complexity because of higher deposition rates and better material quality of WAAM-fabricated parts in comparison to powder-feed AM technology. The overall studies conclude that the multi-wire feed WAAM technique can be utilized to produce functionally graded material (FGMs), alloy–alloy composites (AAC), and various other alloys that are difficult and costlier to fabricate using other manufacturing techniques with the desired composition gradient, suitable mechanical properties, and acceptable oxidation behavior. The chapter well discussed the pros and cons of using a modified version of the conventional WAAM system as a dual-wire feed WAAM technique. The study shows that the use of multiple wires and heat sources decreases the molten pool volume and also increases the deposition rate during fabrication. However, the input of higher heat energy during deposition is responsible for induced residual stress and distortion. Therefore, this brings a crucial step toward the expansion of its applicability areas, opening the door for more demanding applications in the various sector considering the advantages associated with WAAM as a special interest.

REFERENCES

Abdulhameed, Osama, et al. "Additive manufacturing: Challenges, trends, and applications." *Advances in Mechanical Engineering* 11.2 (2019): 1687814018822880. DOI:10.1177/1687814018822880.

Arun, M., and K. Ramachandran. "Effect of welding process on mechanical and metallurgical properties of AA6061 aluminium alloy lap joint." *International Journal of Mechanical Engineering and Research* 5.1 (2015). DOI:ISSN 0973-4562.

Bajaj, Priyanshu, et al. "Steels in additive manufacturing: A review of their microstructure and properties." *Materials Science and Engineering: A* 772 (2020): 138633. DOI:10.1016/j.msea.2019.138633.

Dass, Adrita, and Atieh Moridi. "State of the art in directed energy deposition: From additive manufacturing to materials design." *Coatings* 9.7 (2019): 418. DOI:10.3390/coatings9070418.

Davis, A. E., et al. "Mechanical performance and micro structural characterization of titanium alloy-alloy composites built by wire-arc additive manufacture." *Materials Science and Engineering: A* 765 (2019): 138289. DOI:10.1016/j.msea.2019.138289.

DebRoy, Tarasankar, et al. "Additive manufacturing of metallic components–process, structure, and properties." *Progress in Materials Science* 92 (2018): 112–224. DOI:10.1016/j.pmatsci.2017.10.001.

Dhinakaran, V., et al. "Wire arc additive manufacturing (WAAM) process of nickel based superalloys– A review." *Materials Today: Proceedings* 21 (2020): 920–925. DOI:10.1016/j.matpr.2019.08.159.

Ding, J., et al. "Thermo-mechanical analysis of wire and arc additive layer manufacturing process on large multi-layer parts." *Computational Materials Science* 50.12 (2011): 3315–3322. DOI:10.1016/j.commatsci.2011.06.023.

Dinovitzer, Malcolm, et al. "Effect of wire and arc additive manufacturing (WAAM) process parameters on bead geometry and microstructure." *Additive Manufacturing* 26 (2019): 138–146. DOI:10.1016/j.addma.2018.12.013.

Ezugwu, E. O., Z. M. Wang, and A. R. Machado. "The machinability of nickel-based alloys: A review." *Journal of Materials Processing Technology* 86.1–3 (1999): 1–16. DOI:10.1016/S0924-0136(98)00314-8.

Feng, Yuehai, et al. "The double-wire feed and plasma arc additive manufacturing process for deposition in Cr-Ni stainless steel." *Journal of Materials Processing Technology* 259 (2018): 206–215. DOI:10.1016/j.jmatprotec.2018.04.040.

Geng, Haibin, et al. "Optimisation of interpass temperature and heat input for wire and arc additive manufacturing 5A06 aluminium alloy." *Science and Technology of Welding and Joining* 22.6 (2017): 472–483. DOI:10.1080/13621718.2016.1259031.

Gloria, Antonio, et al. "Alloys for aeronautic applications: State of the art and perspectives." *Metals* 9.6 (2019): 662. DOI:10.3390/met9060662.

Jafari, Davoud, Tom H. J. Vaneker, and Ian Gibson. "Wire and arc additive manufacturing: Opportunities and challenges to control the quality and accuracy of manufactured parts." *Materials & Design* 202 (2021): 109471. DOI:10.1016/j.matdes.2021.109471.

Khan, Anas Ullah, and Yuvraj K. Madhukar. "An economic design and development of the wire arc additive manufacturing setup." *Procedia CIRP* 91 (2020): 182–187. DOI:10.1016/j.procir.2020.02.166.

Kozamernik, Nejc, Drago Bračun, and Damjan Klobčar. "WAAM system with interpass temperature control and forced cooling for near-net-shape printing of small metal components."

The International Journal of Advanced Manufacturing Technology 110.7 (2020): 1955–1968. DOI:10.1007/s00170-020-05958-8.

Kumar, Vishal, et al. "Parametric study and characterization of wire arc additive manufactured steel structures." *The International Journal of Advanced Manufacturing Technology* 115.5 (2021): 1723–1733. DOI:10.1007/s00170-021-07261-6.

Liberini, Mariacira, et al. "Selection of optimal process parameters for wire arc additive manufacturing." *Procedia Cirp* 62 (2017): 470–474. DOI:10.1016/j.procir.2016.06.124.

Martina, Filomeno, et al. "Tandem metal inert gas process for high productivity wire arc additive manufacturing in stainless steel." *Additive Manufacturing* 25 (2019): 545–550. DOI:10.1016/j.addma.2018.11.022.

Nath, Shukantu Dev, and Sabrina Nilufar. "An overview of additive manufacturing of polymers and associated composites." *Polymers* 12.11 (2020): 2719. DOI:10.3390/polym12112719.

Qi, Zewu, et al. "Microstructure and mechanical properties of wire+ arc additively manufactured Al-Mg-Si aluminum alloy." *Materials Letters* 233 (2018): 348–350. DOI:1016/j.matlet.2018.09.048.

Qi, Zewu, et al. "Microstructure and mechanical properties of wire+ arc additively manufactured 2024 aluminum alloy components: As-deposited and post heat-treated." *Journal of Manufacturing Processes* 40 (2019): 27–36. DOI:10.1016/j.jmapro.2019.03.003.

Ramkumar, K. Devendranath, et al. "Investigations on the microstructure, tensile strength and high temperature corrosion behaviour of Inconel 625 and Inconel 718 dissimilar joints." *Journal of Manufacturing Processes* 25 (2017): 306–322. DOI:10.1016/j.jmapro.2016.12.018.

Reisgen, Uwe, Rahul Sharma, and Lukas Oster. "Plasma multiwire technology with alternating wire feed for tailor-made material properties in wire and arc additive manufacturing." *Metals* 9.7 (2019): 745. DOI:10.3390/met9070745.

Ren, Lingling, et al. "Microstructure and properties of Al-6.0 Mg-0.3 Sc alloy deposited by double-wire arc additive manufacturing." *3D Printing and Additive Manufacturing* 9.4 (2021): 301–310. DOI:10.1089/3dp.2020.0039.

Rodrigues, Tiago A., et al. "Current status and perspectives on wire and arc additive manufacturing (WAAM)." *Materials* 12.7 (2019): 1121. DOI:10.3390/ma12071121.

Rosli, Nor Ana, et al. "Review on effect of heat input for wire arc additive manufacturing process." *Journal of Materials Research and Technology* 11 (2021): 2127–2145. DOI:10.1016/j.jmrt.2021.02.002.

Sharma, Ashutosh, Y. S. Shin, and Jae-Pil Jung. "Influence of various additional elements in Al based filler alloys for automotive and brazing industry." *Journal of Welding and Joining* 33.5 (2015): 1–8. DOI: ISSN 2287-8955.

Shen, Chen, et al. "In-situ neutron diffraction study on the high temperature thermal phase evolution of wire-arc additively manufactured Ni53Ti47 binary alloy." *Journal of Alloys and Compounds* 843 (2020): 156020. DOI:10.1016/j.jallcom.2020.156020.

Teja, Kodati, et al. "Optimization of mechanical properties of wire arc additive manufactured specimens using grey-based Taguchi method." *Journal of Critical Reviews* 7.9 (2020): 808–817. ISSN: 2394-5125.

Vaezi, Mohammad, Philipp Drescher, and Hermann Seitz. "Beamless metal additive manufacturing." *Materials* 13.4 (2020): 922. DOI:10.3390/ma13040922.

Vaghani, Minakshi, S. A. Vasanwala, and A. K. Desai "Stainless steel as a structural material: State of review." *International Journal of Engineering Research and Applications* 4 (2014): 657–662. ISSN: 2248-9622.

Wahab, D. A., and A. H. Azman. "Additive manufacturing for repair and restoration in remanufacturing: An overview from object design and systems perspectives." *Processes* 7.11 (2019): 802. DOI:10.3390/pr7110802.

Wang, Jun, et al. "Characterization of wire arc additively manufactured titanium aluminide functionally graded material: Microstructure, mechanical properties and oxidation behaviour." *Materials Science and Engineering: A* 734 (2018): 110–119. DOI:10.1016/j.msea.2018.07.097.

Wang, Jun, et al. "In-situ dual wire arc additive manufacturing of NiTi-coating on Ti6Al4V alloys: Microstructure characterization and mechanical properties." *Surface and Coatings Technology* 386 (2020): 125439. DOI:10.1016/j.surfcoat.2020.125439.

Wang, Yanhu, et al. "In-situ wire-feed additive manufacturing of Cu-Al alloy by addition of silicon." *Applied Surface Science* 487 (2019): 1366–1375. DOI:10.1016/j.apsusc.2019.05.068.

Wu, Bintao, et al. "Effects of heat accumulation on microstructure and mechanical properties of Ti6Al4V alloy deposited by wire arc additive manufacturing." *Additive Manufacturing* 23 (2018a): 151–160. DOI:10.1016/j.addma.2018.08.004.

Wu, Bintao, et al. "A review of the wire arc additive manufacturing of metals: Properties, defects and quality improvement." *Journal of Manufacturing Processes* 35 (2018b): 127–139. DOI:10.1016/j.jmapro.2018.08.001.

Wu, Wei, et al. "Parameters optimization of auxiliary gas process for double-wire SS316L stainless steel arc additive manufacturing." *Metals* 11.2 (2021): 190. DOI:10.3390/met11020190.

Yang, Zhenwen, et al. "Fabrication of multi-element alloys by twin wire arc additive manufacturing combined with in-situ alloying." *Materials Research Letters* 8.12 (2020): 477–482. DOI:10.1080/21663831.2020.1809543.

Yehorov, Yuri, Leandro João da Silva, and Américo Scotti. "Balancing WAAM production costs and wall surface quality through parameter selection: A case study of an Al-Mg5 alloy multilayer-non-oscillated single pass wall." *Journal of Manufacturing and Materials Processing* 3.2 (2019): 32. DOI:10.3390/jmmp3020032.

Yilmaz, Oguzhan, and Adnan A. Ugla. "Shaped metal deposition technique in additive manufacturing: A review." *Proceedings of the Institution of Mechanical Engineers, Part B: Journal of Engineering Manufacture* 230.10 (2016): 1781–1798. DOI:10.1177/0954405416640181.

Yu, Zhanliang, et al. "Microstructure and mechanical properties of Al-Zn-Mg-Cu alloy fabricated by wire+ arc additive manufacturing." *Journal of Manufacturing Processes* 62 (2021): 430–439. DOI:10.1016/j.jmapro.2020.12.045.

Zhai, Yuwei, Diana A. Lados, and Jane L. LaGoy. "Additive manufacturing: Making imagination the major limitation." *Jom* 66.5 (2014): 808–816. DOI:10.1007/s11837-014-0886-2.

Chapter 7

Introduction to the New Emerging Micro-electron Beam Welding Technology

A Sustainable Manufacturing

Anupam Kundu, D.K. Pratihar, Debalay Chakrabarti, and Vidyapati Kumar

7.1 INTRODUCTION

In the last few decades, there had been a rapid growth of advanced welding technologies, such as electron beam welding, laser welding, robotic welding, friction-stir welding, and others. Electron beam welding (EBW) is a well-recognized method, which is capable of joining even high-thickness metals with optimum weld strength. If this technology is utilized for micro-level joining or precision welding, it may be called micro-electron beam welding (micro-EBW). It is becoming popular nowadays, as it provides minimal thermal stress on the part of a micro-element, exact energy feeding, wear-less tool, and the precise control of beam movements (Kundu et al., 2019). Application-wise, it spreads from micro-mechanical fabrication and micro-drilling to the microelectromechanical system (MEMS) and medical science (Kundu et al., 2019; Weglowski et al., 2016). The micro-EBW machine has some extra features that exhibit an enhanced quality of welding. Features-wise, it provides exact heat input, high-level focus ability, minimal energy input, and high-resolution scanning electron microscope (SEM) image for analysis (Dilthey et al., 2006). SEM helps in carrying out both welding and analysis (Smolka, 2005).

This technology is highly preferable for the welding of reactive materials, where the welding materials may form a reaction in an open environment. Micro-EBW is usually carried out under the complete vacuum condition (Smolka et al., 2004; Hwang et al., 2005). So, it is free from environmental hazards like contamination and reactions (Kundu et al., 2019). A few other essential attributes make it more favorable for the micro-systems arena, like possibility of handling extremely tiny beam diameter and yielding low thermal stress, and thus declare this technology superior for the smallest spot or line welds of both metallic and non-metallic substrates (Dilthey et al., 2006; Weman, 2003; Dilthey et al., 2005). This micro-joining technology (i.e., micro-EBW) permits intensely high-frequency oscillation movements, precise energy input, and superior control, which sharpens the quality of micro-welding.

The micro-EBW technology has some specific boundary conditions and requirements to be followed in the area of micro-welding. This technology offers a green environment, due to its processing in a fully CNC (computerized numerical control)-assisted vacuum chamber (Smolka, 2005). This CNC system provides the finest visual quality control and makes it a green welding process for hybrid micro-assembly purposes (Kundu et al., 2019). Generally, welding of small parts or micro-components requires optimal contact conditions. Although these micro-fine sections are highly flexible in nature, the best possible setting is tough to implement (Smolka, 2005; Smolka et al., 2004; Reisgen et al., 2009).

DOI: 10.1201/9781003270027-7

7.2   NEED FOR MICRO-ELECTRON BEAM WELDING

Micro-EBW is carried out in the cleanroom environment with the smaller beam diameters and more accurate energy input compared to the laser. It provides fruitful results for the spot or line joints of the metal and non-metals (Kundu et al., 2019). On the other hand, micro-EBW technology has some key factors that make it a sustainable manufacturing process. It is processed in a clean room environment (vacuum), which yields a green and pollution-free welding process. It eliminates the probability of wastage by not using any filler material. Micro-EBW requires a low-heat input for a small time. It does not provide any negative environmental impact and consumes less energy compared to other manufacturing processes. So, these factors make this new technology a more demanding one for micro-welding. The macro-range EBW machines are not suitable for tiny and thin substrates, as the ranges of their beam power vary from 100 W to a few kilowatt. This range of high power is not ideal for the joining of micro-parts and micro-fabrications (Smolka et al., 2004; Weman, 2003). Another reason for choosing micro-EBW is its in-built attachment of a scanning electron microscope (SEM), which works simultaneously for the observations and necessary analysis. This technology is also found to be much more promising for micro-hybrid assembly. Micro-EBW offers the following benefits:

*   It can efficiently weld both similar and dissimilar (like aluminum to copper, stainless steel to Kovar) material with high precision (Kundu et al., 2019).
*   It can provide high-precision microlevel narrow welding and heat-affected zones (HAZ) (weld width ~150 μm).
*   It has high efficiency of heat utilization (Dilthey et al., 2006).
*   Micro-EBW ensures precise heat control with accurate dosing (Dilthey et al., 2006)
*   The probability of getting welding defects is significantly less (porosity, undercut and overcut, and hot cracking) (Kundu et al., 2019).
*   It can provide deep penetration with less distortion.

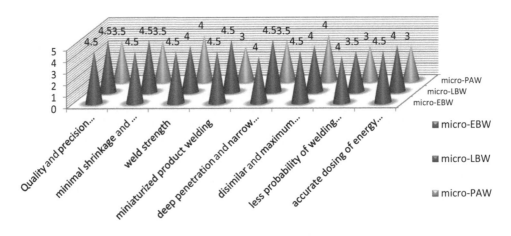

Figure 7.1 3D graphical representation of multiple welding aspects for micro-EBW, micro-LBW, and micro-PAW.

In the present scenario, multiple advanced welding processes are used to fulfill our market demand according to the products' requirements. Some of these techniques are EBW, laser beam welding (LBW), and friction stir welding (FSW), but in the area of advanced micro-welding, micro-EBW, micro-LBW, and some extensions of micro-plasma arc welding (micro-PAW) are used nowadays (Kah et al., 2015). All these welding processes are well efficient in their respective areas of applications in different sectors. Figure 7.1 presents the performances of these micro-welding processes in terms of multiple welding aspects measured on a scale of 0–5 (representing 3.0, 3.5, 4.0, and 4.5 as mean, moderate, moderately high, and very high values, respectively) (Kundu et al., 2021).

7.3 SPECIAL FEATURES OF MICRO-EBW

Micro-EBW machine has several new features, which make it the most advanced and efficient joining process, as described in the following sections.

7.3.1 Pulsed Beam

Usually, a pulsed beam is highly incoherent and time-dependent. The pulses fluctuate over time. The pulsed beams raise the confined heat input by refining the width-to-depth ratio (Weglowski et al., 2016; Kundu et al., 2021; Pratihar et al., 2012).

7.3.2 High-Definition Imaging Facility

It is quite difficult to carry out surface inspection and analysis for a micro-welded part using the standard optical images (Pratihar et al., 2012). By utilizing the high-resolution ultra-modern scanning electron microscope (SEM) present in micro-EBW, very fine surface images measured up to the few micrometer level can be easily revealed and analyzed.

7.3.3 Multi-beam

The multi-beam technology is useful in the area of welding, particularly, micro-welding, where multiple beams of electrons travel across the weld zone through the different positions with very high velocity, and the kinetic energy is being transformed into thermal energy. The material melts, and joining takes place. It reduces welding deformation and undesirable stresses produced during welding. With the help of these techniques, thermal cycles are also modified and that enhances the fabrication quality (Kundu et al., 2021; Fan et al., 2017).

7.4 WORKING PRINCIPLE OF EBW AND MICRO-EBW

Electron beam welding (EBW) is a high-energy fusion welding process. A collimated beam of high-velocity electron strikes the workpiece, the material gets heated, and coalescence is being induced by heating the workpieces (Kundu et al., 2019). The kinetic energy of the electrons is transformed to heat energy or thermal energy, so the workpiece material gets melted, and joining takes place (Kundu et al., 2019; Weglowski et al., 2016). Thus, mechanical energy is being reformed into thermal energy. EBW is usually processed under a complete vacuum condition to prevent the scattering of the electron. Because of an open ambient, there is a tendency for the scattering of an electron due to its hitting with air molecules. As a result, the power density may get reduced (Weglowski et al., 2016; Smolka et al., 2004). EBW is a green welding process and can be efficiently used for delicate assemblies. This process is operated by a CNC-controlled fully automated system for the better quality of fabrication. EBW has several merits: both similar and dissimilar materials (with varying hardness values and melting points starting from soft materials to hard materials) can be joined efficiently (Weglowski et al., 2016; Kundu et al., 2021). No filler material is required during welding. EBW can provide good weld strength with a high depth-to-width ratio with a moderately high weld speed (Jain, 2013).

The gun chamber of an electron beam is integrated with several vital parts, such as an anode, a filament, a grid electrode, magnetic focus lens, deflection coils, and others (Weglowski et al., 2016) (refer to Figure 7.2) (Kumar, 2012). The joined components are protected from the contamination and environmental hazard during the welding cycle, and a high vacuum is maintained (Smolka, 2005; Smolka et al., 2004).

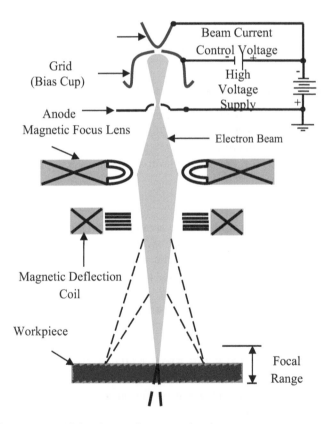

Figure 7.2 A schematic view of the electron beam gun chamber.

Courtesy: Modified from P. Kumar et al. ©NPTEL, 2012. Course: Advanced Manufacturing Processes. URL: https://nptel.ac.in/course/112107077/35

The working principle of micro-EBW is quite similar to that of the EBW, but some modifications have been done for the purposes of better fabrication and analysis. These include the usage of high-resolution SEM and function generators with control software that gives some extra importance to micro-range applications with the precise control (Reisgen et al., 2009). Since the electron beam welding machines operated in the macro-range (with the average beam power of 100 W and up to several kilowatts) are not suitable for the welding of micro-components, less power is used for micro-welding (Dilthey et al., 2006). In micro-welding, a substrate requires the least possible energy input with high resolution and accuracy. Thus, a suitable scanning electron microscope (SEM) is attached to the setup in such a way that it can work for the dual purposes of observation and welding, as shown in Figure 7.3 (Kundu et al., 2019; Reisgen et al., 2009).

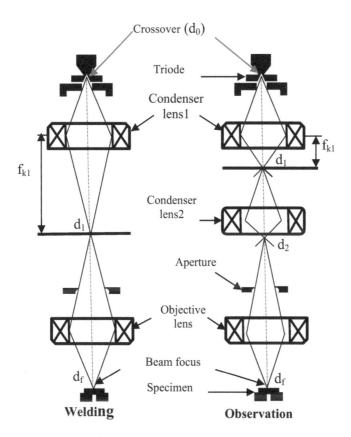

Figure 7.3 A converted SEM representing beam track during welding and observing modes: f_{kl} represents the distance between condenser lens I and crossover I.

Source: Kundu, A., Pratihar, D.K., and Pal, A.R. (2019). A review on micro-electron beam welding with a modernized SEM: process, applications, trends and future prospect, *Journal of Micromanufacturing*, Copyright © 2019, SAGE Publishing.

7.5   CONSTRUCTIONAL DETAILS OF MICRO-EBW MACHINES

Generally, in EBW machines or micro-EBW machines, the basic concept is the same, and most of the machines may be divided into three categories, as given in the following, except in some instances (Schultz, 1993).

- Cycle-type production machines
- Continuous welding machines
- Universal welding type

The micro-EBW machine, including all the aforementioned categories, has some primary components:

- Gun of an electron beam
- Multiple electromagnetic lenses used for positioning and controlling the beam
- Working chamber
- Vacuum system
- Instruments for control and adjustment
- Power supply
- Scanning electron microscope
- Jigs and fixtures arrangement
- Microcomputer interface

From these components, the beam gun chamber and working chamber are the most important ones for a micro-EBW machine. Additionally, a micro-computerized control desk is used for all kinds of data feeding or input and ultraprecise monitoring of the machines (Schultz, 1993). The entire process is run by the PLC-implemented supremely mechanized CNC system (Cohen et al., 2019). The vacuum pump is attached behind the chamber wall, which is acoustically insulated (Kundu et al., 2021).

The electrons are emitted from the electron gun, which is linked with a power supply by some special cables. Next, the vital part is the working chamber, where a complete vacuum condition is maintained, so that the precise welding and joining can take place. A CNC-based flexible table is moved across X and Y directions. For the purpose of visualization, a high-speed camera is also attached.

A micro-EBW machine can be divided into three main sections:

- *Beam generation section:* Here, electrons are triggered from the intensely heated filament and accelerated toward the anode using the high voltage between them (Weman, 2003; Dilthey et al., 2005).
- *Beam manipulation section:* In this part, multiple electromagnetic lenses are used to focus the electron beam toward the weld joint (Weman, 2003; Dilthey et al., 2005).
- *Forming and working chamber:* In this section, a high vacuum is maintained, where the material gets heated, fused, and melted, and the joining takes place.

The normal operating conditions of the micro-EB gun chamber are as follows:

- Accelerating voltage (typically 5–60 kV)
- Beam current (0.05–35 mA)

7.6 MICRO-EBW PROCESS PARAMETERS

The process parameters may be divided into two groups: one is directly related to the micro-electron beam welding, and the other one is relevant to the material features, as discussed subsequently.

The parameters related to the micro-electron beam welding are as follows:

- Accelerating voltage
- Welding speed
- Beam current
- Beam diameter
- Focal distance
- Vacuum level
- Beam power

The parameters relevant to the material features, which may partially affect the welding quality, are as follows:

- Material thickness
- Thermochemical properties of the materials or an alloy
- Type of materials or alloys

The process parameters' range varies depending on the material or alloy properties (like thermal conductivity, thermal diffusivity, melting point, hardness) and welding thickness.

Normally, in the case of micro-EBW, the parameters lie in the range of 5–60 kV for accelerating voltage, 0.05–35 mA for beam current, 1–100 mm/sec for welding speed, and beam diameter of less than 50 μm.

7.7 TECHNOLOGY USED IN MICRO-EBW

A few new technologies are involved in micro-electron beam welding, as discussed in the following sections.

7.7.1 Scanning Electron Microscope

A scanning electron microscope (SEM) is a high-resolution microscope with the better capability of surface image analysis. SEM is mainly concerned with an intense beam of high-energy electrons, which generates an extensive array of signals at the specimen surfaces. The excited electrons are accompanied by a huge amount of kinetic energy, which is scattered and it induces a broad span of signals that are shown in Figure 7.4 (Kundu et al., 2019). The signals produced by SEM are the secondary electron (SE), backscattered electron (BSE), diffracted backscattered electron (EBSD), and X-rays. Out of these, BSE and SE have been mainly used for sample imaging and carrying out its analysis. Backscattered electrons reveal the compositional contrast in the various phases of the sample, and secondary electrons possess surface morphological and topological information about the samples (Kundu et al., 2019). The extracted signals from electron sample interactions bear a detailed information about the sample on crystal structure, microstructure, material orientation, grain boundaries, and chemical composition of the materials. Generally, experimental data are collected from a small portion focused on the sample surface (Weglowski et al., 2016). The SEM range varies from 1 cm to 5 μm in width, with its magnification and resolution varying in the ranges of (20X, 30,000X) and (50 nm, 100 nm), respectively (Kundu et al., 2019; Weglowski et al., 2016).

An SEM has the following components: source of electron beam or gun, electromagnetic lenses, various signal detectors, data output display/monitor, and sample stage.

For carrying out microstructural analysis, SEM is the most vital instrument. Nowadays, it is utilized for micro-welding tools in micro-EBW machines. The SEM works as the dual purposes in a single tool in this machine, namely, welding, and observation (refer to Figure 7.3) (Kundu et al., 2019; Reisgen et al., 2009). Continuous monitoring with the analysis for small substrates or materials may require precise energy input with high-resolution vision (Reisgen et al., 2009). In micro-EBW machines, SEM provides and ensures excellent controllability and high focus ability of the beam that possess the most flexible key demanding sustainable manufacturing technology. Micro-EBW can be ideal for micro-fabrication of the smallest spot or line weld, as the SEM sample image is formed by line scanning of metallic and non-metallic materials.

7.7.2 Function Generator

A function generator is a piece of electronic test equipment that generates several kinds of electrical wave patterns over a wide range of frequencies. It provides a few essential waveforms like a sine wave, triangular wave, and square wave, which are either single-shot or repetitive by nature (Dilthey et al., 2006). A triggering source and an integrated circuit are required to generate a different waveform. In addition, by using the control module, some standard geometries are used like line, circle, and ellipse (Kundu et al., 2019; Dilthey et al., 2006; Reisgen et al., 2008). The function generator can control the random structure of welding and the beam deflection between multiple points at high speed. It works on multiple points of joining, so this process can be called multiple beam technology, which is capable of controlling the output channel, and X and Y deflections. The block diagram is shown in Figure 7.5 (Kundu et al., 2019).

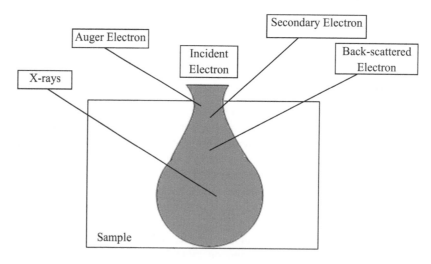

Figure 7.4 Different signals produced while using the SEM.

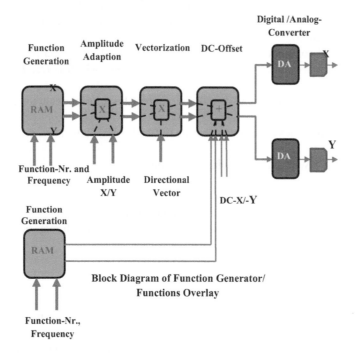

Figure 7.5 A schematic view representing the formation of X/Y deflection and function overlay.

Source: Kundu, A., Pratihar, D.K., & Pal, A. R. (2019). A review on micro-electron beam welding with a modernized SEM: process, applications, trends and future prospect, *Journal of Micromanufacturing*, © 2019, SAGE Publishing.

7.7.3 Control Software

Control software is an electronic control unit (ECU) that controls different electrical systems or subsystems coupled to a function generator (Kundu et al., 2019). Similarly, the figure file is a graphical presentation of different waveform plots made from the data given by the function generator using standard geometries like line, circle, and spline. Figure files are run by ASCII (American Standard Code for Information Interchange) code, and are drawn by either excel or Mathcad. A software module is evolved for these, which interfaces and controls the SEM with a function generator. The application software runs the figure file. Similarly, an SEM image of the sample joint is loaded into the control software. At last, the output data are transformed into an ASCII file that is adaptable for controlling the software of the generator (Kundu et al., 2019; Dilthey et al., 2006).

7.8 ADVANTAGES AND LIMITATIONS OF MICRO-EBW

Micro-EBW has the following advantages:

- Minimal distortion and shrinkage are produced during the welding (Dilthey et al., 2006).
- It can provide precise heat input, which leads to the welding of heat-sensitive elements (Dilthey et al., 2006; Smolka et al., 2004).

- The weld strength of micro-EBW welded parts is almost 95% of the base metal, which is much more than that of any other welding process (Jain, 2013; EBW vs LBW, ebind-ustries.com., 2015).
- Micro-welded components have a lower probability of crack and porosity, as micro-EBW is processed under complete vacuum conditions, which is free from oxides and contaminations.
- There is less possibility of undercut and overcut because of accurate control of heat input or other parameters.
- Micro-EBW provides a narrow weld zone (FZ and HAZ) up to a few micrometers level.

However, micro-EBW has the following limitations:

- Initial setup is costly to make,
- There is a limitation in the size of the vacuum chamber,
- There is a chance of producing X-rays during welding, which may be harmful to the operator (Jain, 2013).

7.9 APPLICATIONS

Innovations of micro-EBW created new probabilities for manufacturing structures at the micrometer and nanometer levels. Using this technique, multiple fabrications can be done, which include optical, magnetic, electronics, and biological devices for computation and control purposes. Micro-EBW provides more compact and miniaturized devices for mechanical to electrical systems or subsystems, which makes a key demand technology for the current trends (Kundu et al., 2019). A pictorial view of various applications of micro-EBW is shown in Figure 7.6. The following are some of the important applications:

- It can fabricate micro-parts or micro-components (Dilthey et al., 2005).
- It can weld any kind of dissimilar material at micrometer levels (Kundu et al., 2021).

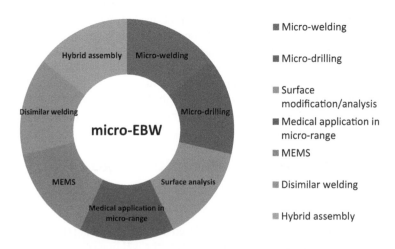

Figure 7.6 A pictorial view of various applications of micro-EBW.

- It can be used for welding of various types of materials.
- It can modify the surface up to a few micrometer levels (Reisgen et al., 2009).
- It can be used for precise micro-drilling (Kundu et al., 2019).
- It has some medical applications requiring high precision.
- It can be used for hybrid assembly.

7.10   DIFFERENT PATTERNS OF JOINING

Various modes of joining, such as single scan, layer scan, and multiple scans are generally used for micro-fabrications. All these three modes are adapted in micro-EBW, as shown in Figure 7.7 (Kundu et al., 2019). The mode of layer scan is primarily used for soldering purposes; similarly, both single and multiple scans are fairly suitable for welding (Dilthey et al., 2006). The mode of a single scan is characterized by the weld path, which is scanned only once. On the other hand, line feed is mainly used for the mode of layer scan (Kundu et al., 2019; Dilthey et al., 2006). In multiple scan mode, the beam oscillates across the weld zone more than once.

7.11   A FEW IMPORTANT TECHNOLOGIES USED IN MICRO-EBW

7.11.1   Microelectromechanical System (MEMS)

Nowadays, MEMS is a key demanding technology that deals with the fabrication of various micro-mechanical and micro-electronics components used in medical science, automotive

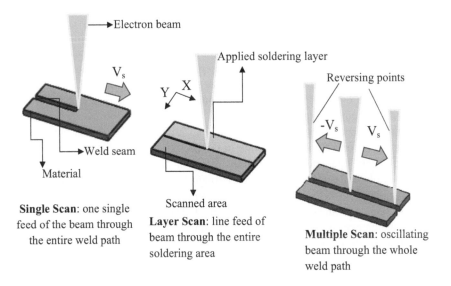

Figure 7.7 Multiple techniques of scanning during welding. (a) Single scan. (b) Layer scan. (c) Multiple scans. V_s represents the velocity of scan.

Source: Kundu, A., Pratihar, D. K., & Pal, A. R. (2019). A review on micro-electron beam welding with a modernized SEM: process, applications, trends and future prospect, *Journal of Micromanufacturing*, © 2019, SAGE Publishing.

industries, and others. In the last decades, micro- and nanotechnology used miniaturization that integrates with microelectronic components, which massively changed our world. A few years ago, mobile phones, computers, and laptops were not common, but these are used frequently nowadays. It has been made possible due to the enormous modification and transformation of microelectronics (Smolka et al., 2004; Dilthey et al., 2005). Microelectronics has gained that much of success because of its integration with miniaturization and MEMS. Miniaturization and functional integration have been used starting from optics to mechanics, where MEMS played a role of an accelerator (Tanasie et al., 2008; Ogawa et al., 2009; Gilleo, 2005).

7.11.2   Carbon Nanotubes

A carbon nanotube (CNT) is a new technology that belongs to the nanosystem technology having better integration of mechanical and electronic properties extracted from its molecular structures (Iijima, 1993; Iijima, 1991). When a single-layer graphite or graphene sheet is rolled and converted into a hollow-type cylinder, it is called carbon nanotubes. The size of this cylinder may be up to a few micrometers in length and a few nanometers in diameter (0.7 nm), and both the ends are sealed with a cap of fullerene. If the carbon tube's wall consists of a sheet of single layer, then it is called single-wall carbon nanotubes. Similarly, whenever this single-layer CNT is combined with van der Walls force, it creates a long rope type of component (Iijima, 1991). In the last decades, CNT had been mainly used in microelectronics, like the fabrication of microelectronics devices, transistors, and micromechanical components (Ajayan et al., 2000).

7.12   CURRENT ISSUES AND CHALLENGES IN MICRO-EBW

In the last few years, micro-EBW had grown exponentially and was established as an emerging technology in different sectors. Multiple complicated fabrication works had been successfully finished to prove the potential of microscale welding. The new ultra-precision SEM imparts a vital role in this success. Though some of the important challenges are still faced during the welding, it is expected to sort out these problems faced by this technology in future, as discussed subsequently.

Beam characterization issues: For carrying out micro-welding, beam characterization is an important part and a novel challenge in the present scenario like the control of beam size, material–beam interaction, and also distortion or deflection. Usually, the micro-parts are highly flexible and due to this, it is very tough to optimize the controlling parameters (Reisgen et al., 2009). Similarly, it is difficult to control the problems related to heat distribution and selection of beam spot size. The heat-affected zones are directly influenced by these parameters, which control the quality of EB welding.

Parametric studies on optimization of weld parameter for multiple alloys: Weld parameters' optimization and control are quite difficult to carry out, in the case of micro-welding, especially for reactive and newly developed smart materials.

Weld joint quality: The quality of a welded product mainly depends on post-fabrication testing. It may be either destructive or non-destructive. In destructive methods, large or prototype samples are generally used for testing, but the same material cannot be used again. In micro-parts welding, non-destructive methods are highly preferred, and material can be reused again. This is a current issue in testing different materials using the NDT (Pratihar et al., 2012). Finally, in situ inspection is highly preferred for micro-welding, if possible, but it is not so easy to carry out.

7.13   SUMMARY AND FUTURE PROSPECTS

This chapter briefly enlightens the various aspects of micro-EBW. Some new techniques are also discussed, which are the integrated part of this emerging technology. The chapter deals with the special features and working principles, favorable applications, advantages, new technologies, and process parameters of micro-EBW. In this chapter, the authors also try to discuss the requirement of micro-EBW, different modes of joining, current challenges, and methods of overcoming these problems.

Over the last few decades, advanced manufacturing technologies were improved very rapidly starting from macro- to micro-manufacturing and some extent to nanotechnology. The innovation in the area of micro-welding or precise fabrication discovers a unique level of track for manufacturing the different structures at micrometer levels. With the help of this innovation, the micro-fabrication process is utilized to join various micro-mechanical to microelectronics components.

Nowadays, materials are welded in various dimensions as per the requirements. It may vary from a few millimeters to some micrometers level. The current age is dependent on new compact technologies. It may be used in high-precision non-traditional manufacturing processes. This high-notch process is mainly preferred on micro-systems, such as micro-cutting or micro-fabrication. The new micro-EBW technology offers some unique features like precise heat control, accurate dosing of energy input, very less distortion, less chance of welding defects (crack, overcut, undercut), and a high rating of accuracy, making this technique ideal for micro-systems, MEMS, micro-mechanical fabrication, and microelectronics. This process is done under complete vacuum conditions, which protect a high degree of contamination, dust, and oxide formation, as a tiny dust particle can influence the welding quality significantly.

There is a huge scope for research and study, in future, on micro-EBW. Some of these scopes for future study are as follows:

- In the case of micro-welding, the computation of weld geometrics is an important problem for future research.
- Beam characterization (beam size, beam material interactions, and distortion) of micro-welding is another critical issue kept open for further study.
- Though micro-welding uses a precise amount of heat, heat distribution, and control, simulation for micro-welding is a good topic for further research.
- The quality of the micro-welding component is essential, so more NDT-based testing and in situ inspection are required to be carried out in the future.
- Implementing this kind of technology in hybrid micro-assembly like microelectronics and MEMS, hybrid micro-mechanical components joining, and fabrication of precise medical implants are also significant research issues.
- Micro-EBW is still in the research stage, and only a few similar and dissimilar materials have been joined efficiently up to a few micrometers level. So, some materials like titanium-based alloys, nickel-based superalloys, aluminum alloys, tungsten alloys, and some composite materials are to be joined through the micro-EBW in future.

ACKNOWLEDGMENT

The first and fourth authors are grateful to the Ministry of Human Resources Development, Government of India, for the financial support they received during their doctoral study.

DECLARATION OF CONFLICTS OF INTEREST

The author(s) declared no potential conflicts of interest with respect to the research, authorship, and/or publication of this chapter.

FUNDING

The author(s) received no financial support for the research, authorship, and/or publication of this chapter.

REFERENCES

Ajayan, P. M., Schadler, L. S., & Giannaris, C. 2000. Single-walled carbon nanotube-polymer composites: Strength and weakness. *Advanced Materials* 12: 750–753.

Cohen, Y., Faccio, M., & Pilati, F. 2019. Design and management of digital manufacturing and assembly systems in the industry 4.0 era. *International Journal of Advanced Manufacturing Technology* 105: 3565–3577. https://doi.org/10.1007/s00170-019-04595-0.

Dilthey, U., Brandenburg, A., & Moller, M. 2005. Joining of miniature components. *Welding and Cutting* 52(7): E143–E148.

Dilthey, U., & Dorfmuller, T. 2006. Micro-electron beam welding. *Microsystems Technologies* 12(7): 626–631.

Electron Beam vs. Laser Beam Welding. 2015. www.ebindustries.com/electron-beam.

Fan, J., Zhang, W., & Qi, B. 2017. Influence of multi-beam electron beam welding technique on the deformation of Ti6Al4V alloy sheet. *Rare Metal Materials and Engineering* 46(9): 2417–2422. https://doi.org/10.1016/s1875-5372(17)30208-4.

Gilleo, K. 2005. *MEMS / MOEM Packaging: Concepts, Designs, Materials, and Processes*. McGraw-Hill, Austin, TX.

Hwang, I., & Na, S.J. 2005. A study on heat source modeling of scanning electron microscopy modified for material processing. *Metallurgical and Materials Transactions* B 36(1): 133–139.

Iijima, S. 1991. Helical microtubules of graphitic carbon. *Nature* 354: 56–58.

Iijima, S. 1993. Single-shell carbon nanotubes of 1 nm diameter. *Nature* 363: 603–605.

Jain, V. K. 2013. *Micromanufacturing Processes*. CRC Press, Taylor & Francis Group, Boca Raton, FL.

Kah, P., Rajan, R., & Martikainen, J. 2015. Investigation of weld defects in friction-stir welding and fusion welding of aluminum alloys. *International Journal of Mechanical and Materials Engineering* 10(1): 26. https://doi.org/10.1186/s40712-015-0053-8.

Kumar, P. 2012. *NPTEL: Mechanical Engineering—Advanced Manufacturing Processes*. NPTEL. https://nptel.ac.in/courses/112/107/112107077/.

Kundu, A., Jaipuria, S., & Pratihar, D. K. 2021. *Electron Beam Welding: Current Trends and Future Scopes*. NOVA Science, Hauppauge, NY. ISBN: 978-1-53619-685-6. https://novapublishers.com/shop/handbook-of-welding-processes-control-and-simulation/

Kundu, A., Pratihar, D. K., & Pal, A. R. 2019. A review on micro-electron beam welding with a modernized SEM: Process, applications, trends, and future prospect. *Journal of Micromanufacturing* 2(2): 220–225. https://doi.org/10.1177/2516598419855186.

Ogawa, H., Yang, M., & Matsumoto, Y. 2009. Welding of metallic foil with an electron beam. *Journal of Solid Mechanics and Materials Engineering* 3(4): 647–655.

Pratihar, D. K., Dey, V., & Bapat, A. V. 2012. *Micromachining: Electron Beams for Macro and Micro-Welding Applications*. Narosa Publishing House Pvt. Ltd., New Delhi, pp. 221–240.

Reisgen, U., & Dorfmuller, T. 2008. Developments in micro-electron beam welding. *Microsystem Technologies* 14(12): 1871–1877.

Reisgen, U., & Dorfmuller, T. 2009. Micro-electron beam welding with a modified scanning electron microscopy: Findings and prospects. *Journal of Vacuum Science and Technology* B 27(3): 1310–1314. https://doi.org/10.1116/1.3137026.

Schultz, H. 1993. *Electron Beam Welding*. 1st ed. Woodhead Publishing, Cambridge, England.

Smolka, G. 2005. Development and characterization of an electron beam plant for the assembly of micro components. *Aachen Reports on Joining Technology*. ISBN: 978-3-8322-4686-0. https://www.shaker.de/de/content/catalogue/index.asp?lang=de&ID=8&ISBN=978-3-8322-4686-0.

Smolka, G., Gillner, A., & Bosse, L. 2004. Micro-electron beam welding and laser machining-potentials of beam welding methods in the micro-system technology. *Microsystem Technologies* 10(3): 187–192. https://doi.org/10.1007/s00542-003-0347-2.

Tanasie, G., Bohm, S., & Bartle, J. 2008. Electron beam machining of micro-system products. *Second European Conference & Exhibition on Integration Issues of Miniaturized Systems-MOMS, MOEMS, ICS, and Electronic Components*, Barcelona, Spain. https://ieeexplore.ieee.org/document/5760593.

Weglowski, M., Błacha, S., & Phillips, A. 2016. Electron beam welding-techniques and trends-review welding of high-carbon nano bainitic steels view project electron beam welding e techniques and trends e review. *Vacuum* 130: 72–92. https://doi.org/10.1016/j.vacuum.2016.05.004.

Weman, K. 2003. *Welding Processes Handbook*. 2nd ed. Woodhead Publishing. www.woodhead-publishing.com.

Chapter 8

Laser Welding of Materials

A Comprehensive Overview

Aparna Duggirala, Bappa Acherjee, and Souren Mitra

8.1 INTRODUCTION

Laser welding has emerged as an indispensable joining method in industries to join different materials with varied thickness based on the respective applications (Ion, 2005). The features like narrow heat affected zone, remote location operations, and automation make laser welding a preferred choice for aerospace, automobile, shipbuilding, and defense applications. The body in white concept of car is manufactured using laser welding due to its precise heat application and ability to weld sheet metals with slender thickness. The concept of laser welding started with the advent of CO_2 laser, the most robust laser for welding till date. Subsequently, the variants like Nd:YAG, semiconductor and diode lasers have made the laser welding more desirable. The keyhole mode welding firmly joins materials as thin as 10 mm in a lesser time. Ease of automation, low-heat-affected zone, and ability to weld variety of metals are the salient advantages of the laser welding. In spite of these advantages, higher initial cost, requirement of skilled labor, and limitation to weld certain materials which are highly reflective limit the widespread use of this technology. The following sections explain the process mechanics, metallurgical features, and defect formation. The aspects of numerical simulation along with the case study are presented at the end.

Chronology of Laser Welding Progress

Different steps in the advancement of laser welding are given in Table 8.1.

Table 8.1 Timelines in the Evolution of Laser Welding (Shin et al., 2020)

Year	Significant Development
1917	Lasers originated with Einstein's 1917 derivation of Plank's law of radiation
1939	Valentin Fabrikant invented the concept of stimulated emission to amplify radiation
1950	The quantum theory of stimulated emission helped to develop the "maser"
1959	A narrow beam of coherent light called "laser" was obtained from optical resonator
1960	The first laser prototype using synthetic ruby with a wavelength of 694.3 nm was developed at the Hughes research laboratory in Malibu, California
1963	Kumar Patel had developed a low-cost, high-efficiency CO_2 laser at AT&T Bell labs in 1963. For more than 50 years, this was the most widely used industrial laser

(Continued)

DOI: 10.1201/9781003270027-8

Table 8.1 (Continued)

Year	Significant Development
1970	New soldering techniques were developed to aid in the advancement of electronic miniaturization
1980	The science of welding was advanced through the use of robots, a variety of gas mixtures, and wide range of electrode materials
1985	Introduction of multi-kilowatt radio frequency exciting units
1988	1 kW Nd:Yag laser came into commercial use
1990s	Lasers progressed from laboratories to industrial units
1997	Rofin-Sinar produced 2 kW diffusion cooled CO_2 machines. Trumpf and Luminoics brought out 4 kW Nd:Yag machines that operated on continuous mode of welding
2000	Magnetic pulse welding was introduced. Diode laser welding is utilized to join thin foils made of steel and titanium
2008	Development of laser-arc-hybrid welding development
2013	Gas metal arc welding-brazing was developed as a procedure to weld the steel used in automobiles
2015	The Ohio State University invented a technique called vaporizing foil actuator welding (VFAW) for joining dissimilar materials. The procedure is comparable to a hybrid of explosive welding and magnetic pulse welding (MPW)
2016–till date	Laser welding is predicted to grow moderately during this period. The market is growing due to increased demand from end-user industries such as automotive, medical, and electronics. In the medical and jewelry industries, laser welding is utilized for limited manual welding, die and tool fabrication, and maintenance

8.2 PROCESS MECHANICS

Laser welding technology uses the thermal and photonic attributes of laser light. Coherent and monochromatic beams with low divergence and high brightness are more effective in converging heat at the weld joint with very little light energy. The high-intensity heat source is used to weld a varied range of materials with thicknesses ranging from a few micrometers to 20 mm (Elijah, 2009). Laser welding produces three different regions on the worked components (Figure 8.1): base material that remains unchanged, fusion zone containing those

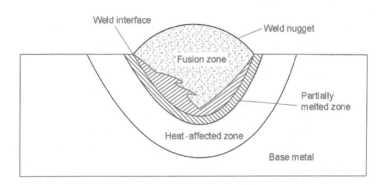

Figure 8.1 Different zones in laser-welded component.

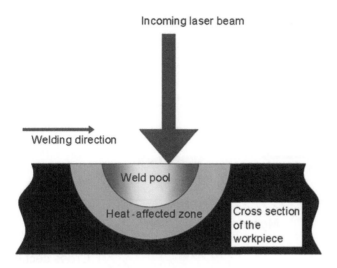

Figure 8.2 Pictorial representation of conduction mode laser welding.

materials that form when the molten pool solidifies followed by a heat-affected zone. This zone is a part of the base material and undergoes considerable change in the material properties due to heat influence. When a laser beam strikes the weldable surface, a considerable percentage of the beam gets reflected. The absorbed beam energy heats the surface, thereby increasing the temperature and creating the molten pool that forms around the welded edges. Further, a portion of the material evaporates due to alloying components and high temperature.

Laser welding mechanics can be classified into two modes: conduction and keyhole configuration (Majumdar and Manna, 2013).

8.2.1   Conduction Mode Laser Welding

This mode occurs when the intensity of the absorbed beam is less than the threshold limit, as shown in Figure 8.2. This is dependent on the absorptivity of the material, scanning speed, and input power. For a majority of metals, an input power higher than 10^5 W/cm^2 initiates the keyhole, below which, the conduction mode welding occurs. The dimensions of the weld pool are determined by the conduction of heat from focus and speed with which the beam traverses. The heat transfer by conduction and the Marangoni surface effect are the predominant mechanisms in this mode of welding (Fabbro, 2013). The resulting heat-affected zone is shallow and wide when compared to the keyhole.

8.2.2   Keyhole Mode Laser Welding

Keyhole mode (Figure 8.3) utilizes the minimum amount of heat possible. Once a laser beam impinges on the surface of the specimen, a significant quantity of heat is reflected away possibly due to metal characteristics, while an amount of radiation absorbed by the surface melts it above the melting point. The molten pool thus formed absorbs more radiation than solids, thereby heating the component even further till the vaporization temperature is obtained. A portion of the

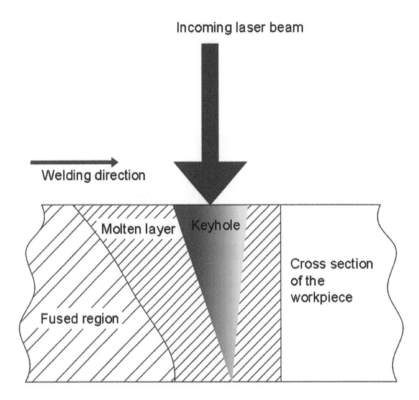

Incoming laser beam

Welding direction

Molten layer   Keyhole

Fused region

Cross section
of the
workpiece

Figure 8.3 Pictorial representation of keyhole welding.

material evaporates, forming a cylindrical passage known as a keyhole. The heat energy is distributed throughout the keyhole, allowing for deep penetration. Due to the relative movement of the weld head and the workpiece, the keyhole passes through the material, forming a weld seam.

8.2.3   Role of Vapor Plume in Laser Welding

The plasma plume surrounds the welding environment along with the work specimen (Figure 8.4). Keyhole welding is accomplished by the high intensity of laser light that produces higher heat extraction rate than the supply rate due to conduction, convection, and back reflection in the metal (Shcheglov et al., 2013). The local heat increases beyond the boiling temperature of the metal resulting in evaporation and dispersion. Further, the recoil pressure forms a keyhole cavity by exerting a piston effect (Semak and Akira, 1997). The metal vapor gets ionized due to its contact with the laser beam. This ionized vapor combines with the surrounding air and shielding gas, forming an electrically neutral plasma inside the keyhole cavity (keyhole plasma) and above the surface of the workpiece (plasma plume) (Coviello et al., 2002). This plasma supports the backscattering, beam absorption, and refraction processes in the keyhole by supplying the necessary energy (Chen et al., 2017; Wang et al., 2006). The quality of the plasma plume depends on the surface purity and the ambient conditions (Xue et al., 2022). High-speed cameras and X-ray imaging techniques are employed to study the plasma plume and its role in keyhole formation.

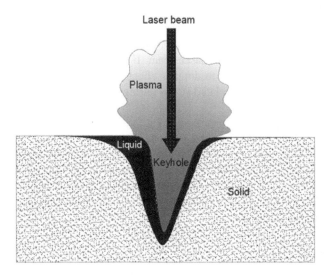

Figure 8.4 Plasma plume surrounding the laser beam and the surface of the component.

8.3 LASERS WELDING SYSTEMS

Appropriate selection of heat source, beam transmission equipment, processing setup equipped with suitable optics, and attachments is necessary to obtain a laser-based joint. Here, a high-intensity beam is focused onto a component, tracing the defined geometry. Though the basic laser variants and the parts employed remain same, certain differences in the components and shielding gas delivery mechanism differentiate laser welding from other processes. The generation and transportation of laser beam to the workpiece for laser welding is achieved through the following.

8.3.1 Laser Generation

The interaction of electromagnetic radiation with the surrounding matter spurs spontaneous emission along with absorption. When the atoms attain their excited state, stimulated emission takes place that initiates a laser light.

8.3.2 Optical Resonator

The main purpose of optical resonator is to define the direction for laser beam. The photons attain their respective states by means of optical resonator.

8.3.3 Laser Pumping

Lasers are often classified by the pumping mechanism also. This mechanism is also called as population inversion and is the main reason for lasing action. The pumping mechanism is broadly divided into optical pumping and electrical pumping and continuous or pulsed pumping.

8.3.4 Broadening Mechanisms

This mechanism influences the profile of the spectroscopic line after the emission of the laser. In homogeneous broadening, every atom in the beam possesses a bandwidth equal to that of the entire bandwidth of the laser beam. If the atoms gain at various frequencies, it is heterogeneous broadening due to doppler effect. Broadening mechanism provides the advantage of reducing the airgap-related issues during butt welding.

8.3.5 Beam Modification

Beam modification is attained by using a focusing lens, Cartesian axis along with shield gas, and filler materials. Lens with small focal number is suitable for high-speed welding of thin materials. Maximum speed is employed, if the focusing depth is equal to the thickness of plates to be welded. Focusing lens with high f number reduce the damage to the lens due to the weld spatter or backscattering.

8.3.6 Beam Delivery

The delivery of the laser beam is frequently accomplished by fiber optics. It is a simple plug-in fixture that has no mirrors attached. It possesses the flexibility to get attached with robots and in factories that are fully automated. An example Nd:YAG laser welding setup is given in Figure 8.5 indicating different components in the machine.

Broadly, lasers are categorized as gas lasers and solid-state lasers. Among gas lasers, He-Ne lasers, CO_2 lasers, and others are in use. Solid-state lasers comprise Nd:YAG laser, diode laser, disk laser, and fiber laser in their category.

8.3.7 Gas Lasers

This group of laser machines use pure gas or a combination of gases to intensify the light. The most popular lasers in this group are helium-neon, argon ion, and carbon dioxide. CO_2 laser welding machines are considered to be the most robust ones and are employed for a varied range of applications, including laser welding, cutting, and other material processing techniques (Tarasov, 1986).

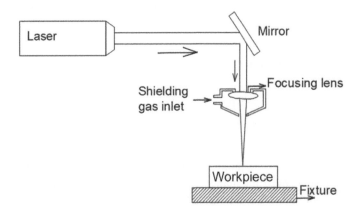

Figure 8.5 Block diagram of a typical Nd:YAG laser setup.

8.3.8   Solid-State Lasers

Also known as doped insulator lasers. The lasers are built around an insulating material that is doped with impurities such as ions. These lasers' primary advantages are their durable build, ease of maintenance, and great efficiency. They are capable of generating large peak powers. Nd:YAG are the most frequently used solid-state lasers. The active medium for these lasers is neodymium (Nd^{3+}). The oxide crystals of yttrium aluminum garnet (YAG) are doped with Nd^{3+} ions to generate lasers under this group. The low excitation threshold and high thermal conductivity generate both pulsed waves and continuous waves. The shorter wavelength is suitable for welding aluminum alloys as they have absorption of 75–90% of the irradiation (Elijah, 2009; Ion, 2005). The pulsed operated Nd:YAG is suitable for micro-welding applications which require precise heat control.

8.3.9   Diode Lasers

The diode lasers operate near-infrared rays at a wavelength in the range of 0.8–1.1 μm. The continuous wave mode of the diode laser is comparable to the He-Ne gas laser. The stacking of diodes improves the power output considerably. These lasers are suitable for laser welding plastics.

8.3.10   Fiber Lasers

These lasers operate at a wavelength of 1.07 μm in the near-infrared region. The power is delivered through an optical fiber cable. The commonly doped elements are erbium (Er), neodymium (Nd), or ytterbium (Yb). The diode laser is used for the pumping process. The lasers have a compact design and the quality of the generated beam is also high.

8.3.11   Disk Lasers

These lasers operate in the near-infrared region wavelength of 1.03 μm. They are doped with ytterbium or yttrium vanadate crystal and are designed to reduce the thermal lensing effect as the heat produced in the disk is eliminated by heat sink placed on the other side of the disk resulting in effective cooling. The beam quality is also good in disk lasers.

8.4   JOINT CONFIGURATION

A variety of joint configurations (Table 8.2) is possible using the laser welding technique. The type of joint is often decided by the demand of the job at hand.

8.5   LASER WELDING PARAMETERS

Laser welding is a complex and multivariable process that is influenced by the properties of welding equipment, laser beam, and workpiece (material and dimensions). The major contributing factors to laser material processing are as follows:

i.   **Laser power:** This is the most critical aspect in the creation of keyholes. Aluminum alloys require larger power densities for welding due to the material's physical features such as greater reflection and thermal conductivity. Additionally, imperfections such as splatter, undercut, underfill, and dropout are typical at higher powers (Adisa et al., 2019).

Table 8.2 Different Types of Joint Configurations that Are Welded Using Laser Welding Mechanism

Joint name	Figure	Preparation
Butt Joint		The two plates are placed close to each other without any air gap
Lap Joint		Welding is done along the overlapping edges by placing the components one upon the other
Lap Joint		Welding occurs when laser is moved along the edges of the plates that are placed one upon the other
T-Joint		The components are placed one upon the other in T-shape and welded along the edges
T-Joint		Beam moves along the edge of the component after placing them one upon the reverse T- shape

ii. **Welding speed:** Welding speed affects the penetration depth. When welding at a faster rate, keyhole's depth is reduced and the weld geometry retains its shape. When the welding speed is reduced, conduction mode welding occurs. As scanning speed is inversely proportional to input power for a unit length of the seam, the depth and width of the weld decrease with speed.

iii. **Laser wavelength**: When incoming laser light strikes a material's surface, it is absorbed based on the wavelength, to perform welding or cutting. The longer wavelength of light is more reflecting, which limits its application to alloys with high reflectivity. A shorter wavelength beam improves process stability, resulting in increased Fresnel absorptivity.

iv. **Laser modes**: The energy distribution in the beam affects the final shape of the weldment. A TEM00 mode has a higher d/w ratio compared to other modes of higher grade. This is due to the smaller beam radius which gives out a highly focused beam (Ayoola et al., 2019). However, multimode beams with a bigger diameter are suitable for butt welding as it reduces the need for stringent edge fit-out.

v. **Laser beam polarization**: In any type of laser material processing, the polarization aspect of a laser beam has a significant influence on the outcome. The direction of polarization of the beam in the plane of interaction is referred to as p-polarization, whereas its orthogonal direction is referred to as s-polarization. The quantity of light that is reflected or transmitted varies depending on the incident polarization. The weld pool behavior is directly impacted by energy accumulation, masking phenomena, and beam polarization. The effect of beam polarization on the welding process is material- and process-dependent (Courtois et al., 2013).

vi. **Laser spot size and power density:** The obtainable power density is decided by the spot size. When the spot size is smaller, the power density is higher leading to narrow welds without proper fusion.

vii. **Beam divergence**: The expansion and propagation of a beam are termed beam divergence. A beam with low divergence allows the concentration of energy in a small area helping formation of keyhole.

viii. **Focal length:** Along with waist diameter and depth of focus, focal length determines spatter generation, spot size, and the level of damage to optics. The depth of penetration and the width of the weld pool is affected by the focal length.

ix. **Depth of focus (DOF):** The maximum deviation that the component can make from the position of optimal focus while maintaining a reasonable weld quality is DOF. When welding thicker pieces, a greater DOF is advantageous.

x. **Position of the focal plane:** The smallest waist diameter is generally at the focal plane. Positive focusing (above the material's surface) produces more plasma, which results in less energy reaching the material's surface.

xi. **Shielding gas:** Shielding gas hinders oxidation of weld seam, reduces instabilities of keyhole, and prevents vitiation of the seam from moisture, grease, and other contaminants, along with protecting the optics. Argon and helium are the frequently employed gases for laser welding. Nitrogen is an excellent material for fabricating keyholes and deep penetration welds but creates nitrides that weakens the weldment. In comparison to Ar and Ni, He performs better as a shielding gas under the given operating parameters and processing conditions (Xu et al., 2019).

xii. **Surface quality:** The components to be welded ought to be freed from impurities, as local melting of contamination results in poor metallurgy and rapid loss of material from the surface during the welding process.

8.6 METALLURGICAL ASPECTS OF LASER WELDING

The phase transition and microstructure of the material are caused by the flow of heat during the welding process. The resulting microstructure affects the weldment's characteristics, the presence of residual stresses, defects, and distortion. Three distinct zones are formed according to the extent to which heat affects a specific area during the laser welding process: the fusion zone, the partially melted zone, and the heat-affected zone (Kou, 2003). Generally, the fusion area of laser-welded components has a fine grain microstructure.

8.6.1 Fusion Zone (FZ)

This is the zone in which actual melting occurs. Melting and solidification occur here to form the weld joint. This melting and solidification process is material- and structure-dependent. For instance, the cracking phenomenon in aluminum alloys can be minimized by utilizing a silicon-containing filler material as it impacts the material structure (Lippold, 2015). The fusion zone in autogenous welding is a melted and resolidified base material. The only change in the microstructure happens as a result of the alloying components evaporation. The composition of the material solidifying in the fusion zone is not as consistent as that of the base material. Grain development in the FZ impacts solidification and is epitaxial (Figure 8.6). In laser beam welding, the grains parallel to the fusion line serve as the nucleation substrate. Due to the proximity of the melt pool liquid to the underlying layer, nucleation occurs easily (Milewski et al., 1999).

The following process variables affect grain development and microstructure:

When the traverse speed of the beam is high, fine grain size results in a refined dendritic structure due to the low-heat input in the case of aluminum alloys, as reported by Guitterez et al. (1996). It is reported that pulsed-mode laser welding results in finer inter-dendritic phases on aluminum alloy (AA) 2024 when welded with Nd:YAG machine. When compared

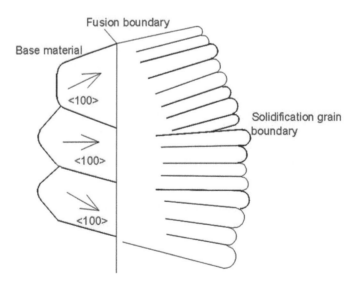

Figure 8.6 Epitaxial growth of the grains in a metal.

to high-duty cycles, low-duty cycles provide a faster cooling rate. When welding at high rates, the columnar grain development is normal to the welding direction (Cao et al., 2007; Gaumann and Kurz, 1998). It is important to consider grain growth while assessing the ductility of components. When the scanning speed is high, Nd:YAG lasers generate weldments with poor ductility (Guitterez et al., 1996).

8.6.2 Unmixed Zone

The unmixed zone (UZ) is present in all fusion weldments. This is a liquid layer that is stationary and thrives when the fluid velocity approaches zero around the fusion line. Evaporation of the liquid metal from the fusion region of contamination on the surface, filler material selection, and process parameters all contribute to the formation of the unmixed region. This region's microstructure is identical to that of the fusion region. The unmixed zone may not be present in high-speed welding processes such as laser beam welding (Lippold, 2015).

8.6.3 Partially Melted Zone

Partially melted zone (PMZ) is the zone between liquid and solid. This zone exists between the material's liquidus and solidus temperatures. Certain variables affect the PMZ range: the alloying elements or impurities in the base metal increase the base metal's actual melting temperature range. The PMZ of aluminum alloys is extremely narrow, about the size of a single grain or two grains (Cao et al., 2007).

8.6.4 Heat-Affected Zone

Heat-affected zone (HAZ) separates the PMZ from the unaffected base metal and it is not subject to melting or evaporation. However, the heating and cooling phenomena, and the input process parameters, affect the microstructure of this region. The microstructure of HAZ is influenced by metallurgical reactions such as recrystallization, phase transitions, precipitate formation, residual stresses, and grain development. The heat-affected zone is quite small when welding at low temperatures or in materials with a high thermal diffusivity. In comparison to other welding procedures, laser welding has the benefit of providing a narrow heat-affected zone.

8.7 DEFECTS IN LASER WELDING

Various types of defects occur in laser-welded components depending on material composition, welding conditions, and so on. These imperfections are categorized into three types: (i) geometrical defects, (ii) internal defects, and (iii) property defects.

8.7.1 Geometrical Defects

8.7.1.1 Deformation or Distortion

The thermal cycles like heating, melting, evaporation, solidification, and cooling that occur in the metal during laser welding lead to change in volume due to thermal expansion or contraction. This leads to distortion or the presence of residual stress and the formation of cracks in the joints. Though distortion is smaller in laser welding due to narrow weld bead, high speed and deep penetration are also responsible for distortion. The larger coefficient of thermal expansion of aluminum leads to distortion in the laser-welded components.

8.7.1.2   Poor Surface Appearance of the Laser Welds

These are generally caused by oxidation, ultrafine particles, fumes, or spatter. Pits and under-cuts also render poor surface appearance.

In laser welding, the metallic silver color of certain metals like aluminum or titanium alloys may change to heavy blue, purple, black, golden color, etc., based on the level of oxidation.

8.7.1.3   Burn through or Meltdown

This defect is said to occur when the molten metal drops down during welding, to form a concave top surface while the bottom surface is convex shaped.

This phenomenon generally occurs in the welding of thin sheets and also when thick plates are welded at high power. The meltdown readily occurs in materials with lower surface tension.

8.7.1.4   Undercutting

The notch that forms along the toe of the weld is called an undercut (Figure 8.7). This defect generally occurs due to high pressure of assist gas, wide weld bead that is formed due to high power density, supply of ample amount of shielding gas from the front side of the component. The effect of gravity also helps in the formation of the undercut.

8.7.1.5   Underfilling

Underfills occur as concave surfaces on the weld bead. This happens during the welding of butt joints with wide gaps and a shortage of filler rods.

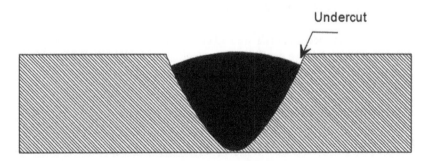

Figure 8.7 Pictorial representation of the undercut.

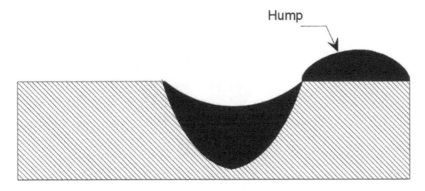

Figure 8.8 Pictorial representation of humping.

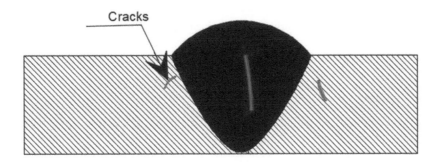

Figure 8.9 Appearance of cracks on welded components.

8.7.1.6 Humping

The development of lumps of weld metal on the bead surface at some intervals is called humping (Figure 8.8). This occurs in narrow weld bead produced at high welding speed and small focused beam, low-vacuum welding. The higher surface tension of the molten pool and backward flow of melt expulsion by plume ejection are the main reasons for humping.

8.7.2 Invisible Defects

8.7.2.1 Hot Cracking

High-temperature cracking on weld fusion zone and heat-affected zone (HAZ) is called solidification cracking and liquation cracking, respectively. This generally occurs along the grain boundary. Causes for solidification cracking may be due to the formation of low solidification temperature liquid film, along the grain boundary.

8.7.2.2 Cold Cracking

This type of cracking occurs below 300°C. It is also known as delayed cracking or hydrogen-assisted cracking. A typical cracking on the weldments is represented in Figure 8.9.

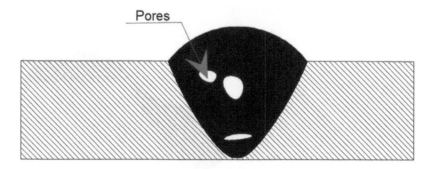

Figure 8.10 Pictorial representation of pores in the weld zone.

8.7.2.3 Porosity

This is the most common defect during laser welding (Figure 8.10). The bubbles that generate at the tip of the keyhole during deep penetration stay as pores in the fusion zone of the weld. Welding conditions, surface cleanliness, and the type of shielding gas affect the formation of pores.

8.7.2.4 Spiking

Saw-like penetration with a periodical variation that forms near the bottom part of a partially penetrated weld bead is called spiking.

8.7.2.5 Swelling of Lap Joint

This type of defect takes place in thin aluminum alloy when it is subjected to low welding speed and high heat input. When a large amount of welding takes place in the upper sheet than in the lower sheet, swelling occurs.

8.7.2.6 Incomplete Penetration

This type of defect occurs in the following ways:

i.   When the weld bead does not reach the bottom of the components
ii. When weld beads that form from both sides do not fuse
iii. When weld forms like a bridge across two pieces

Incomplete penetration (Figure 8.11) occurs in materials that possess high reflectivity and high conductivity like aluminum and copper.

8.7.2.7 Inclusions

Inclusions are said to occur due to the presence of oxides or nitrides during welding. The oxide layer is present on the surface of the aluminum components protecting them in corrosive applications. Nitrides can form in the melt pool when nitrogen is used as a shielding gas. Nitrides present in the molten pool of aluminum 6xxx series protect it from solidification

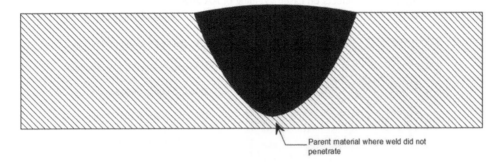

Figure 8.11 Incomplete penetration of the weld seam.

cracking. Though the presence of oxides and nitrides is useful, they will stay as inclusions in the final weld bead and thereby weakening it.

8.7.2.8   *Evaporation Loss of Alloying Elements*

The impurities present on the surface; volatile elements with lower vaporization temperature than the base material when present as alloying elements evaporate quickly leading to defects.

8.7.3   Quality or Property Defects

8.7.3.1   *Reduction in Mechanical Properties*

Certain welding defects like porosity and cracking degrade the mechanical properties of the welded joint. Experimental investigation on certain aluminum alloys of the 5xxx series indicated that the strengths obtained from the welded joints are significantly lower than that of the parent material due to the presence of porosity cracks, underfills.

8.7.3.2   *Degradation in Chemical Properties*

Stress-corrosion cracking and grain boundary corrosion may occur in the HAZ of certain metals and alloys. Improved chemical properties require a better choice of material.

8.7.4   Analysis of Defects in Laser Welding of AA 2024

Table 8.3 presents the micrographs of defects that occurred during laser welding of aluminum alloy 2024 in butt joint configuration. All the experiments are carried out on the pulsed-mode Nd:YAG machine with argon as the shielding gas.

8.8   ANALYTICAL STUDIES IN LASER WELDING

The laser physics is governed by the principles of optics and waves. The entire phenomenon in the laser irradiation zone is highly coupled and not easily decipherable. Numerical simulation of such large models is done with some assumptions that ease the computation and

Table 8.3 Micrographs and Defects on Laser Welding of AA 2024

Micro-graph	Feature description
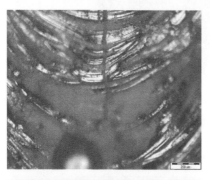	The cracked weld joint developed at average power: 4 kW, scanning speed: 2mm/s, frequency: 8 Hz, and pulse width: 7 ms. The crack propagation is uneven, indicating hot tearing. The crack is 49.68 μm in width
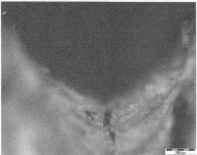	Undercut that occurred at high power and duty cycle. The loss of the molten material due to high power and formation of keyhole are the reasons for this defect
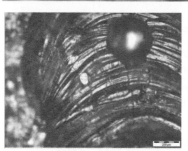	Pores occurred due to the entrapment of the gases and the metal vapor. The use of shielding gas, argon has minimized the pores in these experiments
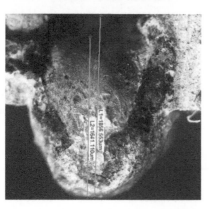	Root drop occurred when welding at high scanning speed and high power and frequency

understanding. Thermomechanical simulations that simulate the impact of welding on the mechanical behavior of the components are termed as multi-physical models. They evaluate the thermal field along with the residual stress and distortion on the large structure and check for the tolerance limits also. While simulating them, shell elements are used that mimic the material and physical conditions of the original structures. The study outlines are reduced and volumetric heat source is employed to ease the computation. The step-by-step mechanical and thermal simulation process is shown in Figure 8.12. Different heat sources (Figure 8.13) are employed to depict the laser beam and its intensity. Double ellipsoidal heat source is proposed by Goldak and is generally used to determine energy transport. Conical heat source provides faster thermal diffusion but FZ is not considered. Combined conical and cylindrical heat source gives the nail shape to the weld seam.

Meshing divides the large structure into small finite elements. Application of loads and boundary conditions is precise with appropriate mesh size. Once boundary conditions are defined according to the practical application of the weldment, mechanical stress and distortion are calculated in the next step. The governing equations consider the system to be in static equilibrium. The plastic deformation is also considered in situations where there is change in

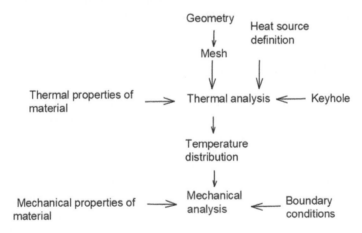

Figure 8.12 Step-by-step process of finite element analysis.

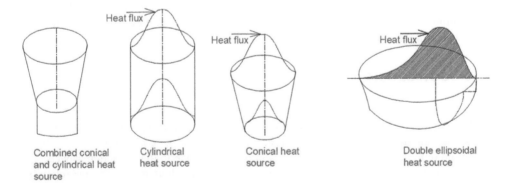

Figure 8.13 Heat sources used in laser welding simulation.

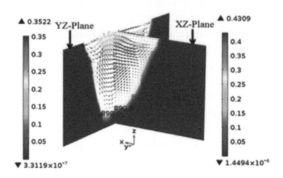

Figure 8.14 Fluid flow pattern in the weld zone obtained from numerical simulation.

volume. Metallurgical models are developed to understand the phase distribution around the weldment. This helps in predicting the zonal behavior of welded component under different operating conditions. Computational fluid dynamics throw a light on the flow pattern of the molten fluid in the weld zone. A typical flow field along with velocity is shown in Figure 8.14.

8.9   LASER MICRO-WELDING

Micro-welding using lasers is gaining prominence in the fabrication of electronic and medical equipment wherein highly precise and no contact workpieces are vital. The components produced by laser welding are found to have better surface attributes compared to those produced by resistance welding. However, typical defects like lack of fusion, distortion, and cracking occur in laser-welded ultrathin materials. Micro-welding using lasers have similar processing parameters and principles as macro-laser welding, albeit the component size is very small and thinner. The laser can be focused into an extremely small beam radii on the order of tens of micrometers. But the high amount of heat generated during the welding process creates a HAZ that sometimes affects the entire component in the case of micro-welding given the smaller dimensions of the component. The problem of micro-welding is still more challenging in the case of light metal alloys like aluminum as high energy density is required initially to overcome the reflectivity of the components. Managing the defects like humping, cracks, and pores is a big challenge that needs to be addressed to obtain sound joints in micro-welding.

8.10   CONCLUSIONS

Laser welding is occupying an indispensable position in joining components due to its versatility and merits. Industries like automobile, aerospace, packaging, shipbuilding, to name but a few, employ laser-welded joints in their products. But there is still a long way to go. The rapidity in the use of lasers still needs improvement due to the economic aspects. The defects that occur during the welding process pose concerns about the quality and reliability of the final joints. Though the industry of lasers is advancing with the arrival of femtosecond and other precise lasers, the use of lasers in welding is still moderate. It is expected to grow in similar terms in near future also. An insight into the experimental and analytical aspects of laser welding is given by discussing the different modes in laser welding, types of welding machines, and different parameters to be adjusted while welding. The welding of aluminum alloy and the limitations in obtaining a successful weld out of AA2024 are discussed briefly. Hence research toward enhancing the applications and reducing the defects show a promising future for laser welding applications.

ACKNOWLEDGMENT

AD thanks Director, School of Laser Science and Engineering, Jadavpur University, for providing the necessary infrastructure in carrying out this work.

REFERENCES

Adisa, Saheed B., Irina Loginova, Asmaa Khalil, and Alexey Solonin. 2019. "Effect of laser welding process parameters and filler metals on the weldability and the mechanical properties of AA7020 aluminium alloy." *Journal of Manufacturing and Materials Processing* 2: 33.

Ayoola, W.A., W.J. Suder, and S.W. Williams. 2019. "Effect of beam shape and spatial energy distribution on weld bead geometry in conduction welding." *Optics and Laser Technology* 117: 280–287. https://doi.org/10.1016/j.optlastec.2019.04.025.

Cao, X., W. Wallac, J.-P. Immarigeon, and C. Poon. 2007. "Research and progress in laser welding of wrought aluminum alloys. II. Metallurgical microstructures, defects, and mechanical properties." *Materials and Manufacturing Processes* 18(1): 23–49. https://doi.org/10.1081/AMP-120017587.

Chen, Y., Z.K. Lei, R.X. Bai, Y.P. Wei, and W. Tao. 2017. "Study on elastoplastic crack propagation behavior of laser-welded 6061 aluminum alloy using digital image correlation method." *Materials Science and Engineering* 281: 012040. https://doi.org/10.1088/1757-899X/281/1/012040.

Courtois, Mickael, Muriel Carin, Philippe Le Masson, Sadok Gaied, and Mikhaël Blabane. 2013. "A new approach to compute multi-reflections of laser beam in a keyhole for heat transfer and fluid flow modelling in laser welding." *Journal of Physics D: Applied Physics* 14: 505305. https://doi.org/10.1088/0022-3727/46/50/505305.

Coviello, Donato, Antonio D'Angola, and Donato Sorgente. 2002. "Numerical study on the influence of the plasma properties on the keyhole geometry in laser beam welding." *Frontiers in Physics* 9: 754672. https://doi.org/10.3389/fphy.2021.754672.

Elijah, Kannatey-Asibu, Jr. 2009. *Principles of Laser Materials Processing*. John Wiley & Sons, Hoboken, NJ.

Fabbro, Remmy. 2013. *Developments in Nd: YAG Laser Welding. Handbook of Laser Welding Technologies*. Woodhead Publishing, Oxford, pp. 47–72.

Gaumann, M., and W. Kurz. 1998. "Why is it so difficult to produce an equiaxed microstructure during welding." *Mathematical Modelling of Welding Phenomenon* 695(4): 125–136.

Guitterez, L.A., G. Neye, and E. Zschech. 1996. "Microstructure, hardness profile and tensile strength in welds of AA6013-T6 extrusions." *Welding Journal* 75(4): 115S–121S.

Ion, John C. 2005. *Laser Processing of Engineering Materials: Principles, Procedures and Industrial Application*. Elsevier Butterworth-Heinemann, Burlington.

Kou, Sindo. 2003. *Welding Metallurgy*. John Wiley & Sons, Hoboken, NJ.

Lippold, John C. 2015. *Welding Metallurgy and Weldability*. John Wiley & Sons, Hoboken, NJ.

Majumdar, Jyotsna Dutta, and Indranil Manna. 2013. "Laser-assisted fabrication of materials." In *Springer Series in Materials Science*, edited by Jyotsna Dutta Majumdar and Indranil Manna, 159–201. Springer-Verlag, Berlin, Heidelberg.

Milewski, J.O., G.K. Lewis, and J.E. Wittig. 1999. "Microstructural evaluation of low and high duty cycle Nd:YAG laser beam welds in 2024-T3 aluminum." *Welding Journal* 72(7): 341S–346S.

Semak, Vladimir, and Akira Matsunawa. 1997. "The role of recoil pressure in energy balance during laser materials processing." *Journal of Physics D: Applied Physics* 30: 2541–2552.

Shcheglov, P. Yu, I.B. Gornushkin, and V.N. Petrovskiy. 2013. "Vapor–plasma plume investigation during high-power fiber laser welding." *Laser Physics* 23: 016001 (7pp).

Shin, Yung C, Wu Benxin, Lei Shuting, Gary J. Cheng, and Y. Lawrence Yao. 2020. "Overview of laser applications in manufacturing and materials processing in recent years." *Journal of Manufacturing Science and Engineering* 142: 110818-1.

Tarasov, L. 1986. *Laser Physics and Applications*. Mir Publishers, Moscow.

Wang, Hong, Yaowu Shi, and Shuili Gong. 2006. "Numerical simulation of laser keyhole welding processes based on control volume methods." *Journal of Physics D: Applied Physics* 39: 4722–4730. http://doi.org/10.1088/0022-3727/39/21/032.

Xu, Jie, Yi Luo, Liang Zhu, Jingtao Han, Chengyang Zhang, and Dong Chen. 2019. "Effect of shielding gas on the plasma plume in pulsed laser welding." *Measurement* 134: 25–32. https://doi.org/10.1016/j.measurement.2018.10.0470263-2241/Ó 2018.

Xue, Boce, Baohua Chang, and Dong Du. 2022. "Monitoring of high-speed Laser welding process based on vapour plume." *Optics and Laser Technology* 147: 107649. http://doi.org/10.1016/j.optlastec.2021.107649.

Chapter 9

IoT-enabled Condition Monitoring and Intelligent Maintenance System for Machine

R. Mohanraj, R. Rajamani, S. Elangovan, S. Pratheesh Kumar, T.G. Sreekanth, C.S. Ramshankar, and R. Sugumar

9.1 INTRODUCTION

Long-term physical engagement with a small number of employees who are highly trained to manage big production equipment is common in manufacturing facilities. Those individuals, on the other hand, are exposed to health and safety risks as a result of poorly maintained facilities and equipment. Maintenance is critical to maintaining a healthy environment, both locally within the institution and globally. Through good maintenance and asset management procedures, waste generated by industrial processes may be reduced and managed. People's health and safety in the production plant may be significantly enhanced and maintained with well-planned and managed maintenance. In comparison to other manufacturing industries, the issues related to health and safety are extremely important since it is a primary source of environmental hazards. This necessitates a plan and comprehensive knowledge of maintenance as a component of a larger system that operates in concert to benefit the entire business. Figure 9.1 shows the importance of maintenance in a business.

Maintenance is the heart of every worldwide enterprise's production system. The output of the production system, such as quality, quantity, and safety, is critical to the enterprise's success. In industry, e-technologies such as online and wireless technologies expand the

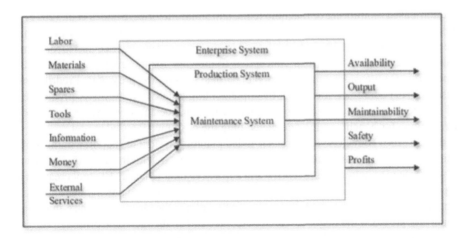

Figure 9.1 Input and output model of the enterprise.

DOI: 10.1201/9781003270027-9

possibilities for combining data from many sources and processing enormous amounts of data to perform sophisticated reasoning and decision-making in order to carry out cooperative operations. The maintenance operator may respond to any crisis quickly and prepare any invention optimally, thanks to real-time monitoring of machine status and configurable notifications. Furthermore, high-rate communications allow for the rapid acquisition of many expertise as well as the acceleration of the feedback response in the local loop, which is connected by the product, the monitoring agent, and the maintenance system. It offers nearly limitless potential for reducing the difficulty of standard maintenance recommendations based on decision-making and product condition assessments.

The smart factory is a method used by companies to integrate everything inside the sector, bringing the virtual and physical worlds together. The IoT is a significant technique behind the smart manufacturing idea. The smart factory initiatives like Industrial Internet Consortium (IIC) in the United States, Smart Manufacturing Leadership Coalition (SMLC) in the United States, Industry 4.0 in Germany, and E-factory in Japan are the major ones across the world. The big data analytics is getting importance as the bulk of data available through the IoT can be meaningfully interpreted with user-defined algorithms for increase in profit of the factory. The smart factory approach is applied to various types of industries such as automobile, manufacturing, chemical, and food supply industries.

The Internet of Things describes environment where network connectivity and processing capabilities are extended to objects, sensors, and ordinary things, allowing them to create, share, and consume data with minimum human interaction. The Internet of Things is getting closer to becoming a reality because of recent technological and business trends. These include pervasive connection, broad IP-based networking adoption, computer economics, miniaturization, data analytics advancements, and the advent of cloud computing. For a discrete manufacturing industry, the connected factory system removes the existing centralized system of activities and replaces it with a mix up of centralized and decentralized system. This improves communication within the industry during the manufacturing process. Also, the system reduces the wastage of materials and energy consumption during the operation so that the entire process could get optimized. This results in improved efficiency of processes across the industry.

A wide range of research articles are available related to predictive maintenance system, condition-based monitoring of machine tools, and smart factories. Simoes et al. (2011) studied the issues related to the different facets of manufacturing organizational performance. Several challenges related to industrial performance measuring practice and theory are recognized and examined. Finally, a conceptual framework for the evolution of manufacturing performance measurements in an organizational setting was presented. Fredriksson and Larsson (2012) addressed the issues of implementing reactive, preventive, predictive, and total productive maintenance for Volvo trucks. Maintenance analysis was performed, which is a technique for benchmarking and developing the customer-focused model.

Ashayeri (2007) developed a computer-aided planning system to support a number of production facilities, each of which is equipped with a CNC machining centers. The planning tool was created to help with preventive and corrective maintenance. The tool creates a good and practical preventive maintenance plan, which allows increasing the number of preventive maintenance tasks completed. Muller at al. (2008) outlined the ideas within the e-maintenance domain and provided the major research outcomes and challenges. The primary criteria that allow for the development of new maintenance support systems were investigated. The study presented a state-of-the-art overview of current industrial and academic research. Liao and

Lee (2010) proposed an approach for developing a reconfigurable prognostics platform that could be used to monitor and anticipate the performance of machine tools in a simple and effective manner. Reconfigurable prognostic can be implemented on the equipment and has the prognostic ability to turn data into performance information. Two industrial cases were used to validate the effectiveness to apply the reconfigurable prognostics platform to various machine health assessment and performance prediction applications.

Mori and Fujishima (2013) developed remote monitoring and maintenance system for machine tool manufacturers. The system relied on a mobile phone unit as a means of communication between the client and the machine tool manufacturer in order to quickly materialize and install the system at the customer's location. More than 10,000 machine tools have been fitted using this technology. The increase in service efficiency was shown in a real-world setting. Radziwon et al. (2014) reviewed the smart technology and discussed the smart factory concept. The challenges of the potential smart factory applications in small and medium enterprises (SMEs) were discussed and a future research outlook was proposed in order to further develop the smart factory concept. Tedeschi et al. (2015) developed the groundwork for the development of a secure remote monitoring system for machine tools that uses a cloud environment to facilitate communication between the client and the maintenance service provider. Cloud technology made it possible to obtain maintenance data from a variety of devices, including tablets and smart phones.

Srivastava (2013) demonstrated condition-based preventive maintenance for a computer numerical control (CNC) turning machine. The methodology of preventive maintenance approach is adopted to determine the preventive maintenance index function, by taking values from a CNC machine. The time needed between the two preventive maintenance visits is determined, which in turn helps in reducing the frequent breakdowns occurring in the CNC machine. Edrington et al. (2014) introduced a web-based machine monitoring system that performs data collection, analysis, and machine event notification for MTConnect compatible machines. The suggested system gave shop managers the information they needed to improve shop floor process efficiency and overall equipment effectiveness (OEE). Lee et al. (2015) discussed current trends toward implementing cyber-physical systems in the manufacturing industry. The 5C architecture was explained, which is capable of automating and centralizing data processing, health evaluation, and prognostics. The architecture encompasses all required procedures, including data acquisition, processing, presentation to users, and decision assistance.

Wei et al. (2016) developed an IoT-based energy management system for industrial clients that included communication architecture and a common information model to make the creation of a demand response (DR) energy management system easier. Singh et al. (2013) developed a system that automates original equipment effectiveness (OEE) of a machine. The software was designed and developed in Visual Basic which calculated OEE of the machine with set of data obtained from the machine. The parameters considered for user interface design are OEE, number of defects, asset utilization, quantity produced, and quality. As a result, a system was developed which calculated OEE with a set of data, where it analyzed the data using the software to identify the losses and measures could be taken to eliminate the problems affecting OEE.

The research articles have provided a good insight into topics like remote monitoring, Internet of Things, smart factory concept, maintenance performance indicators, and prognostic method. With the development of information and communication technologies, highly automated manufacturing delays and unplanned downtime due to unexpected machine breakdown can be avoided.

Maintenance expenses are estimated to account for 1–25% of overall operating costs in manufacturing enterprises (Simoes et al., 2011). Furthermore, it has been stated that around 30% of maintenance expenses are attributed to wasteful spending as a result of poor planning, overtime, and unmet preventive maintenance requirements (Fredriksson and Larsson, 2012). Organizations are increasingly considering maintenance as a vital element of their business, making it more impossible to neglect it. However, recent trends suggest that many production systems are not working as planned. Inadequate maintenance can lead to more unexpected asset failures, which come with a slew of expenses for the company, including delayed production, rework, scrap, labor, replacement parts, late order fines, and lost orders due to unsatisfied customers.

For processing the signals from IoT gateway or from data cloud, IoT software was used. For developing IoT platform for the intelligent maintenance system, IoT software platform which is a web application was used. It contains the most complete set of integrated IoT-specific capabilities. The software can be used to develop front end for the users to view or monitor the data in useful forms and back end to perform the program coding for interpreting the data meaningfully. These situation demands a smart maintenance system which is able to minimize the manufacturing delays and unplanned downtime due to unexpected failure of machines. This work aims to analyze the predictive maintenance of the shop floor machines with respect to their real-time data collection through the sensors.

Present research focuses on how to implement the sensors to collect the various process parameter data and to design a User Interface (UI) using ThingWorx platform. Augmented Reality application for condition monitoring system will be developed to improve rapid response to process upsets or alarm conditions which in turn support equipment maintenance. The proposed predictive maintenance system is a smart maintenance system based on Internet of Things strategy which helps data collection, storage, transmission, analysis, and decision-making. It assists in timely delivery of data regarding the predictive maintenance activities to maintenance crew of a factory and in turn helps in effective utilization of machines and manpower, making the production cost-effective. The history of data stored can be further utilized for data analytics and machine learning which can enhance the capabilities of the shop floor machines.

9.2   TECHNIQUES AND METHODOLOGY

The methodology for the development of predictive maintenance of machine with respect to production forecasting system is given in Figure 9.2. Ineffectiveness of the maintenance system without proper utilization of resources was identified by preliminary analysis which needs to be enhanced for better production. System design and architecture has been proposed for effective maintenance management system. For testing the concept, horizontal turning center was identified. Preventive maintenance checklist has been generated for the production models.

For creating user interface dashboard, IIoT software Thingworx was used. For collecting the various sensor data from the machine, the Kepware gateway system was implemented.

Predictive maintenance system was developed to provide maintenance activities for an effective utilization of machines and manpower. It involves designing, integrating, testing, and validating the system and discussing how the adopted maintenance strategy will improve the factory's operation and performance. The system design covers principle of approach and proposal of the architecture. The system integration and testing involves the formation of test procedures in system level.

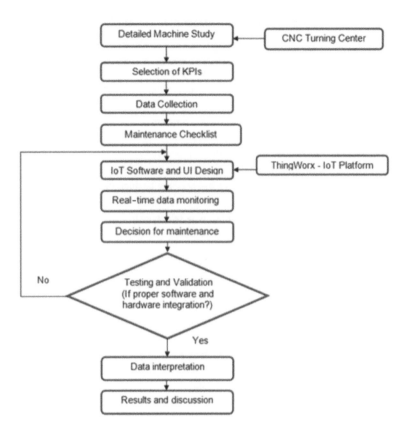

Figure 9.2 Flowchart of the maintenance system.

9.3 INTELLIGENT MAINTENANCE SYSTEM

An intelligent maintenance system (IMS) is a system that is required in all machinery or system to predict and prevent the failure. The failure of a machine or a component leads to the loss of money invested. In order to avoid them, the smart and intelligent system is required to notify the abnormal activity to take a preventive action at a correct time. The advanced sensors are mounted in the system or machinery to collect the current time data, this data is used to analyze and interpret to provide the result. The system is the prognostic activity to avoid the unnecessary wastage of money in the system.

In this case, horizontal turning center in which Smarturn's maintenance activity was planned to implement through the intelligent maintenance system (IMS). For that the machine's preventive maintenance checklist was collected based on its periodic manner.

9.3.1 Ticketing System for Autonomous Maintenance

The maintenance checkpoints are classified based on the stations and frequency of maintenance such as daily, weekly, quarterly, half yearly, and yearly; the checklist is shown in Tables 9.1, 9.2, 9.3, 9.4, and 9.5, respectively.

Table 9.1 Daily Maintenance Checkpoints for Turning Center

S. No.	Maintenance Checkpoint
1	Clean the chips inside the machine
2	Lubrication pump oil level checking
3	Air pressure: 5 kg/cm^2
4	Lubrication unit pressure: 15 kg/cm^2
5	Spindle taper bore cleaning
6	Moisture content in the compressed air
7	FRL unit water draining
8	Hydraulic unit pressure—counter-balance (65 bar)
9	APC unit hydraulic pressure: 15 kg/cm^2
10	Air-conditioning functioning (electrical)
11	Exhaust fan (spindle drive duct) (electrical)

Table 9.2 Weekly Maintenance Checkpoints for Turning Center

S. No	Maintenance Checkpoints
1	Lubricator oil level in FRL unit
2	Cleaning of air filter in AC unit
3	Air filter and fan of electrical cabinet cleaning
4	Coolant tank/filter cleaning
5	Tool pot and tool shank cleaning
6	Check the magazine tool gripper tightness
7	Air filter of spindle oil cooler cleaning (op)
8	Charger tank oil level

Table 9.3 Quarterly Maintenance Checkpoints for Turning Center

S. No	Maintenance Checkpoints
1	ATC magazine cleaning
2	Lubrication pump suction filter cleaning
3	Cracks in hydraulic hoses
4	Oil flow in hydraulic hoses
5	Cleaning of machine lamp
6	APC hydraulic power pack oil level checking
7	Chip conveyor oil level checking

Table 9.4 Half Yearly Maintenance Checkpoints for Turning Center

S. No	Maintenance Checkpoints
1	Slide seal adjustment
2	Check spindle belt tension
3	Slide seal cleaning

(Continued)

Table 9.4 (Continued)

S. No	Maintenance Checkpoints
4	FRL unit cleaning
5	Check the lubrication distributor oil flow
6	Servo motor connector tightness
7	Hydraulic power pack oil replacement
8	Proximity switch, limit switch, and dog checking
9	Electrical control panel cleaning
10	Cable breakage/damage shielding relay terminal/screw tightness
11	Position coder connector tightness

Table 9.5 Yearly Maintenance Checkpoints for Turning Centre

S. No	Maintenance Checkpoints
1	Check battery (system) voltage
2	Cleaning of PCB's cleaning of all proximity/limit switches
3	Cable routing check
4	Replacement of fan filters in control panel
5	Electrical loose connections
6	Slide deal replacement
7	Table level reading

Figure 9.3 Intelligent Maintenance System (IMS) android user interface.

In this ticketing system, the preventive maintenance checklists are added into the mobile device. The operator should enter the checklist in the smart manner: if any check is not in a proper manner, the system will ask to write a remark; otherwise, it will perform and take a snap for that particular check. Figure 9.3 illustrates the IMS smart application.

The IMS system is developed for the machines available in the factory, so in the query page it will ask the employee to sign in their own ID and password. The system also asks the query regarding the company's cell name, department name, machine number, and preventive maintenance period. The final report will be forwarded to the superior officer assigned in the platform of PTC ThingWorx.

9.3.2   Algorithm for Ticket Generation

A machine has a preventive maintenance checklist for which the checks are to be conducted on daily, weekly, quarterly, half yearly, and yearly basis. A checklist is often a logically organized set of action items or criteria that allows the user to track the presence or absence of each item on the list to ensure that they are all evaluated or finished.

The maintenance team is comprised of maintenance operators assigned to set of machines and they perform these maintenance activities based on the schedule. Each checkpoint is an activity that the maintenance operator has to perform. The entire follow-up of these preventive maintenance activities are called the ticketing system.

If the maintenance operator observes any fault against any checkpoint, then it is called as an abnormality. Once an abnormality identified, a ticket will be generated. A ticket is a mechanism used in the factory to track the detection, reporting, and resolution of maintenance problem or an issue of equipment or machine. The status of a ticket can be open, in-progress, closed, and verified. Open ticket means that the issue has not yet been looked into, in-progress ticket means that the maintenance activity has been taken up and is currently in the state of progressing, closed ticket means that the maintenance activity of the problem of particular machine or equipment has been completed by the maintenance team, and verified ticket means that the completed maintenance activity has been verified by the higher authority of the respective department.

The creation of various tickets and its workflow are shown in Figure 9.4. Count of the tickets is used for showing various maintenance metrics, which can be viewed on the IMS

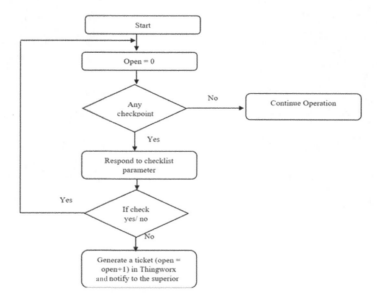

Figure 9.4 Flow chart of ticketing system function.

dashboard. The parameters are status of abnormalities, aging of abnormalities, user-based abnormality status, distribution of abnormalities, and deferred maintenance.

9.4   DASHBOARD DESIGN

The dashboard of Intelligent Maintenance System was designed using software for developing IoT platform, which is web-based software for Internet of Things and has specific development tools and capabilities. The dashboard preparation includes thing templates, things, thing shapes, streams, properties, services, and the mashup. The logics are written in the services in the specified JSON (Java Script Object Notation) format.

The ticketing system was implemented in the dashboard analytics. The main dashboard view of the system is shown in Figure 9.5.

The checkpoints for a machine are checked in the mobile application by the operator and sent to URL in the form of string. Test data for one month has been generated and has been sent to IoT software through Postman. Postman is a Google Chrome application which provides Graphical User Interface (GUI) for constructing request and reading responses. The IoT software services get executed and tickets are generated against each abnormalities. The analytics shows on the dashboard will automatically get updated based on the duration at which it is refreshed.

9.5   SYSTEM ARCHITECTURE

In this research, a preventive maintenance approach is integrated into a machine monitoring framework. The framework gathers data from maintenance operator inputs through a mobile application. The preventive maintenance checkpoints are available in the database of ThingWorx which is web-based software for internet on ThingWorx which

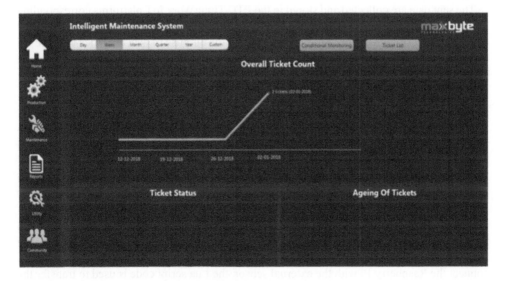

Figure 9.5 Main dashboard view of intelligent maintenance system.

includes preparing thing templates, things, thing shapes, steams, properties, services, and the mashup.

When the operator selects a machine through mobile, the preventive maintenance check-points will appear one by one on the mobile for which the operator has to update with the status of checks, the data pertaining to status of machines are processed in ThingWorx services in which the business logics are written. A unique ticket will be generated for each abnormality. An autogenerated alert mail will be sent to maintenance person. The processed data are displayed on the dashboard as various maintenance metrics such as the abnormalities trend, cumulative abnormalities trend, and aging of abnormalities. These metrics are based on the counts of open, in-progress, closed, and verified tickets. The metrics can also be viewed based on machine, cell, department, or the entire factory. The display of data for time-based metrics such as status of abnormalities can be filtered as daily, weekly, quarterly, half yearly, and yearly. The maintenance team can log on to the dashboard for follow-up of maintenance activities. The maintenance team can log on to the dashboard for the follow-up of maintenance activities. The maintenance work can be assigned based on the workload for each operator and the ticket status can be updated each time during the progress work.

9.6   CONDITION MONITORING

Condition monitoring is the process of collecting and visualizing the raw data into the useful information. For raw data collection process in the CNC turning center, basically two types of methods are followed. In the first method, the data are directly collected from the Program Logic Controller (PLC) with the aid of KepserverEx gateway. Each parametric data has a certain register number in the PLC, so the data can be retrieved from the PLC with the application of KepserverEx system through the ethernet cable. Another method for data collection is initiated by Raspberry Pi or Arduino microcontroller from the external sensors. The collected data are send to the PTC ThingWorx cloud platform with the aid of corresponding things application key.

The sent data are collected and stored in the PTC ThingWorx IoT platforms corresponding thing. The PTC ThingWorx platform is used to interpret the raw data into the useable visual data. The PTC ThingWorx platform is also used for the purpose of data analytics with the aid of various machine learning algorithms.

9.7   SENSOR INTEGRATION

Raspberry Pi is a microcontroller which is used to interface the external sensor data to the cloud platform. It has a special operating system called Raspbian, it has the processor speed range of 700 MHz to 1.2 GHz for the Pi 3, onboard memory range of 256 MB to 1 GB RAM. The external secured memory card is used to store and transfer the data. It also contains the operating system of the Raspberry Pi microcontroller.

Figure 9.6 illustrates the interfacing diagram of Raspberry Pi with the MPU6050 accelerometer sensor. The data which come from the sensor can be visible in python as well as the PTC ThingWorx platform. Two types of codes are used to collect and transfer the data from the machine to the PTC ThingWorx IoT platform. Namely, the python code is used to connect the Raspberry Pi with the external sensor, the Lua script code is used to transfer the data from Raspberry Pi to the ThingWorx platform.

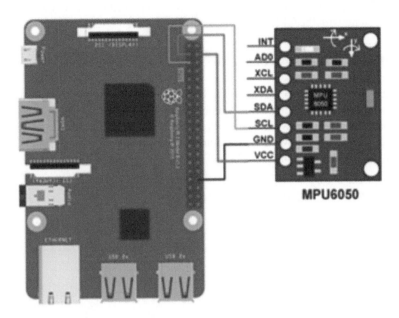

Figure 9.6 Interfacing Raspberry Pi with MPU6050 sensor.

9.8 CONDITION MONITORING OF OVERALL EQUIPMENT EFFECTIVENESS

Overall equipment effectiveness (OEE) is a most effective way to increase the efficiency of production rate and it is determined in Equation (9.1). OEE is the widely accepted tool used in the industry to measure and estimate the productivity of a system. OEE belongs into three measurement terms that are known as availability, performance, and quality. Overall equipment effectiveness (OEE) encourages production companies to step up their manufacturing process and in terms of efficiency, effectiveness, and quality of excellence.

$$\text{Overall equipment effectiveness (OEE)} = (\text{availability} \times \text{performance} \times \text{quality}) \quad (9.1)$$

In the CNC turning center, there is no separate provision for measuring quality of the product. So, the quality value is assumed as 100%. In the availability, the planned production time is equal to the total time when the emergency button is switched off and machine is in on condition. The run time is equal to the total time when the processing light is blowing. The ideal cycle time will differ for each and every product.

9.9 DASHBOARD DESIGN FOR OEE MONITORING

The real-time data collected from the CNC turning center was transmitted to the PTC Thing-Worx platform, from there the user interface is allowed to visualize the processed data which can be easily interpreted by the user.

Figure 9.7 represents the dashboard for the CNC turning center availability, performance, and quality along with the OEE.

9.10   PREDICTIVE MAINTENANCE

Predictive maintenance in the CNC turning center has been monitored by considering various parameters when the machine is in running condition. The surface roughness of the work material has been predicted with the aid of feed rate, spindle speed, and depth of cut during machining. The parameters were selected and experiment study was performed as per L9 orthogonal array design, as shown in Table 9.6.

9.11   EXTERNAL SENSOR MOUNTING

To measure the vibration during the turning process, the MPU6050 vibration sensor was mounted in the tool post of the CNC turning center. Figure 9.8 illustrates the mounting of the MPU6050 sensor in the tool post of the CNC machine.

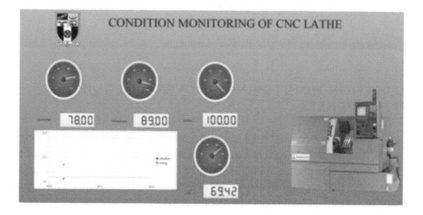

Figure 9.7 Dashboard for Overall Equipment Effectiveness.

Table 9.6 L9 Orthogonal Array for Experimental Study

Test No.	Spindle Speed (rpm)	Feed Rate (mm/rev)	Depth of Cut (mm)
1	1,000	0.1	0.25
2	1,000	0.15	0.2
3	1,000	0.2	0.1
4	1,200	0.1	0.2
5	1,200	0.15	0.1
6	1,200	0.2	0.25
7	1,400	0.1	0.1
8	1,400	0.15	0.25
9	1,400	0.2	0.2

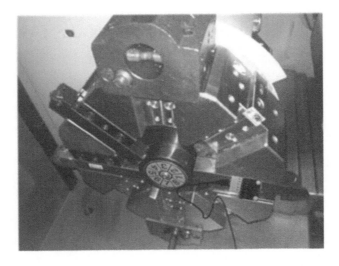

Figure 9.8 Sensor arrangement in the turning centre.

Tungsten carbide material was selected for single-point cutting tool insert, the mild steel workpiece material was selected for experimentation which has the diameter of 50 mm and length of 170 mm. L9 orthogonal array and its parametric condition have been experimented and corresponding values from Program Logic Controller (PLC) and the MPU6050 sensor have been transferred to the PTC ThingWorx platform. The surface roughness value for each and every experimental value was noted.

9.12   DASHBOARD DESIGN FOR PREDICTION MODEL

The dashboard design contains the necessary values to predict the surface roughness and it also includes the tool life calculation parameters. So the user interface consists the feed rate, depth of cut, spindle speed, acceleration in X, Y, and Z directions, prediction value of surface roughness, and tool life value. In addition, the time series graph which illustrates the fluctuation in the values of each parameter with respect to time period is also available.

Figure 9.9 illustrates the dashboard design for the prediction model with the aid of Thing templates, Things, thing Shapes, Streams, Properties, Services, and the Mashup in the PTC ThingWorx IoT platform.

9.13   AUGMENTED REALITY DESIGN

A direct or indirect live view of a physical, real-world environment whose elements are enhanced by computer-generated perceptual information, ideally across several sensory modalities, is referred to as augmented reality (AR). UNITY3D software was used for making the Io-based augmented reality application. The image target–based augmented reality has been created which portraits the various parametric data from the PTC ThingWorx platform. A 3D Creo model diagram is also included in the augmented reality application.

Figure 9.10 illustrates the image target–based augmented reality application for the CNC turning center for visualizing the real-time data from the remote place.

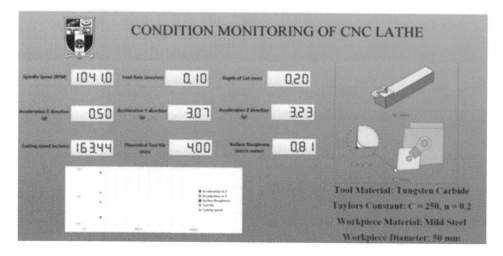

Figure 9.9 Dashboard for surface roughness prediction.

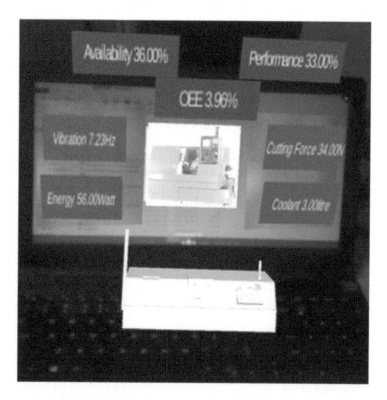

Figure 9.10 Augmented reality application design.

9.14 CONCLUSIONS

Predictive maintenance of shop floor machines with respect to the production forecasting system has been developed by proposing the system architecture. A new system called ticketing system which executes the maintenance through identification, work planning, scheduling, and execution has been designed for Intelligent Maintenance System (IMS). The mobile application for checklists ensures each checkpoint which can be reviewed by the operator without skipping automatically or intentionally. The proposed system is capable of following the preventive maintenance activities in effective manner by providing the right information at right time through analytical calculations on designed dashboard.

The condition monitoring system of CNC turning center has been developed for real-time monitoring the OEE of the machine from the remote place. The prediction model is generated for surface roughness prediction with respect to the machine's feed rate, depth of cut, and spindle speed with the aid of regression analysis.

Augmented reality application created for visualizing the virtual environment with the application of image target–based system. The proposed method will reduce downtime, increase productivity, enhance customer satisfaction, reduce total maintenance cost, and very importantly it is very useful for maintenance scheduling process in mass production industry.

REFERENCES

Ashayeri, J (2007), Development of computer-aided maintenance resources planning (CAMRP): A case of multiple CNC machines centers, *Robotics and Computer-Integrated Manufacturing*, 23:6, 614–623.

Edrington, Ben, Zhao, B, Hansel, A, Mori, M and Fujishima, M (2014), Machine monitoring system based on MTConnect technology, *Procedia CIRP*, 22, 92–97.

Fredriksson, G and Larsson, H (2012), *An analysis of maintenance strategies and development of a model for strategy formulation: A case study*, Goteborg: Chalmers University of Technology.

Lee, J, Ardakani, HD, Yang, S and Bagheri, B (2015), Industrial big data analytics and cyber-Physical systems for future maintenance & service innovation, *Procedia CIRP*, 38, 3–7.

Liao, L and Lee, J (2010), Design of a reconfigurable prognostics platform for machine tools, *Expert Systems with Applications*, 37:1, 240–252.

Mori, M and Fujishima, M (2013), Remote monitoring and maintenance system for CNC machine tools, *Procedia CIRP*, 12, 7–12.

Muller, A, Marquez, AC and Lung, B (2008), On the concept of e-maintenance: Review and current research, *Reliability Engineering and System Safety*, 93:8, 1165–1187.

Radziwon, A, Bilberg, A, Bogers, M and Madsen, ES (2014), The smart factory: Exploring adaptive and flexible manufacturing solutions, *Procedia Engineering*, 69, 1184–1190.

Simoes, JM, Gomes, CF and Yasin, MM (2011), A literature review of maintenance performance measurement: A conceptual framework and directions for future research, *Journal of Quality in Maintenance Engineering*, 17:2, 116–137.

Singh, R, Shah, DB, Gohil, AM and Shah, MH (2013), Overall Equipment Effectiveness (OEE) calculation-automation through hardware and software development, *Procedia Engineering*, 51, 579–584.

Srivastava, M (2013), Condition-based maintenance of CNC turning machine, *International Journal of Mechanical Engineering and Robotics Research*, 2:3, 420–426.

Tedeschi, S, Mehnen, J, Tapoglou, N and RajKumar, R (2015), Security aspects in Cloud based condition monitoring of machine tools, *Procedia CIRP*, 38, 47–52.

Wei, M, Hong, SH and Alam, M (2016), An IoT-based energy-management platform for industrial facilities, *Applied Energy*, 164, 607–619.

Index

3D printing 51, 79–81, 84, 85, 89, 91, 101–103

autonomous maintenance 167

buildup 4, 6, 9, 19, 20

carbon nanotubes 138
condition monitoring 166, 172, 173, 177

directed energy deposition 46, 91

electrode gap 29, 30

fused deposition 65, 80, 81, 85

heat-affected zone 110, 128, 138, 143, 145, 152, 153, 155
hierarchy 79, 80, 84, 85, 89
hybrid additive manufacturing 41, 45, 50, 57, 58, 59

intelligent maintenance system 166, 167, 171, 177
IoT 163–166, 171, 172, 175

mechanistic 16, 18
MEMS 1, 127, 137–139
metallic glass 10, 11, 13, 16, 18, 21, 23
micro-features 1, 2, 3, 13, 29, 30
micromachining 1, 2, 5, 9, 11, 15, 19, 20, 29, 30
microstructure 7, 10, 11, 15, 47, 50–53, 56, 63, 67–69, 94, 96, 97, 99, 103, 106, 110–112, 115, 118, 119, 122, 123, 133, 152, 153
micro-tools 3, 7, 24, 29, 33, 35, 36, 37, 38
minimum uncut chip thickness (MUCT) 6, 7, 14

networking 164

orthogonal array 30, 31, 174, 175

selective laser melting (SLM) 64, 68, 69, 70, 71
size effect 2, 5, 6, 9, 10, 18, 21
stereolithography (SLA) 81, 85–89

ultrasonic 1, 44, 49, 54–56

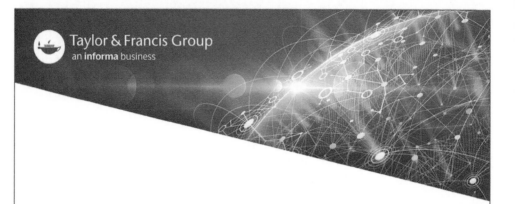